UNEP YEAR BOOK
EMERGING ISSUES IN OUR GLOBAL ENVIRONMENT
2012

United Nations Environment Programme

In Memoriam

Credit: Harvey Croze

Professor Wangari Maathai

Credit: Brigitte Lacombe

Professor Wangari Maathai passed away on 25 September 2011 in Nairobi, Kenya. Professor Maathai was a champion for the environment, human rights and the empowerment of women. Her Green Belt Movement encouraged rural Kenyan women to plant trees in order to improve their livelihoods and curb the effects of deforestation.

Professor Maathai was the first African woman and first environmentalist to receive the Nobel Peace Prize. Honoured for her commitment to women's empowerment and environmental sustainability, she also served in Kenya's Parliament and was appointed assistant minister for environment and natural resources. Wangari Maathai received her doctoral degree from the University of Nairobi in 1971, making her the first woman in Central and East Africa to receive such a degree.

1940–2011

Table of Contents

Preface	v
Acronyms	vi
Executive Summary	vii
Year in Review	**1**
Climate change, extreme weather events and disaster risk management	1
Population dynamics and resource scarcity	4
Energy and climate change mitigation	6
Global biodiversity conservation	11
Looking ahead	12
2011 At a glance	14
2012 Calendar of events	15
References	16
The Benefits of Soil Carbon	**19**
Carbon storage and other vital soil ecosystem services	21
What determines the global distribution of soil carbon?	22
Modelling, measuring and monitoring	23
The vulnerability of soil carbon stocks to human activities	25
Consequences of soil carbon loss and the potential for soil carbon gain	27
The way forward: managing soil carbon for multiple benefits	29
References	32
Closing and Decommissioning Nuclear Power Reactors	**35**
What is nuclear decommissioning?	35
State and trends in nuclear decommissioning	36
Three approaches to decommissioning	37
The challenges of decommissioning	39
Risks associated with decommissioning	43
Lessons learned	47
References	48
Key Environmental Indicators	**51**
Depletion of the ozone layer	51
Climate change	53
Natural resource use	55
Chemicals and waste	61
Environmental governance	61
Looking ahead	62
References	64
Acknowledgements	66
Questionnaire	68

Preface

The 2012 UNEP Year Book spotlights two emerging issues that underline the challenges but also the choices nations need to consider to deliver a sustainable 21st century—urgently improved management of the world's soils and the decommissioning of nuclear reactors.

Superficially they may seem separate and unconnected issues. But both go to the heart of several fundamental questions: how the world will feed and fuel itself while combating climate change and handling hazardous wastes.

The thin skin of soil on the Earth's surface is often one of those forgotten ecosystems but it is among the most important to the future survival of humanity.

The top one metre of soil sustains agriculture, supports forests, grasslands and meadows which in turn generate the conditions for the health and viability of many of the globe's plant and animal species. The top one metre also stores three times more carbon than is contained in the atmosphere. Yet land use change is triggering dramatic losses of soils and the nutrients and carbon stored. The Year Book notes that in some places, soil erosion is occurring at rates 100 times faster than soil is naturally made. More intelligent and integrated policies are needed to reverse these trends.

The Year Book cites no-till policies being pursued in some countries, using illustrative case studies from Argentina and Brazil, that are assisting to store soil carbon with other wide-ranging benefits. It also highlights a pioneering form of agriculture called 'paludiculture' that allows farmers to cultivate rather than degrade peatlands in ways that maintain their enormous carbon stocks while producing crops for sustainable biofuels.

Decommissioning of nuclear power stations is spotlighted as an emerging issue because of the large number of reactors that have ended or are nearing the end of their lives. Close to 140 nuclear power reactors in nearly 20 countries have been closed but only around 17 have been decommissioned and more closures of older plants are scheduled over the coming years and decades. Meanwhile the tsunami that hit the Fukushima nuclear plant in Japan in 2011 has prompted some countries to review their nuclear power programmes.

The Year Book looks at the options and the complexities of decommissioning. It also analyzes another issue for which there remains sparse information, namely the price of making the plants and associated radioactive materials safe for current and future generations. By some estimates decommissioning of a nuclear power plant may cost between 10 per cent and 60 per cent of the initial construction costs–an issue that perhaps needs to be more clearly factored in when energy choices are made along with environmental and social parameters.

This year's Year Book comes in advance of the Rio+20 Summit where governments will reconvene to debate and devise more decisive and accelerated action towards implementing sustainable development and realizing an inclusive Green Economy.

Sound and impartial science is at the core of UNEP's work across all its sub-programmes from climate change and ecosystems to resource efficiency and disasters and conflicts. It will be the foundation upon which nations can act to realize their post Rio+20 aims and aspirations as it has been increasingly for nearly 20 years.

Achim Steiner

United Nations Under-Secretary-General and Executive Director,
United Nations Environment Programme

Acronyms

A/R	afforestation and reforestation
CBD	Convention on Biological Diversity
CFC	chlorofluorocarbon
CH_4	methane
CITES	Convention on International Trade in Endangered Species of Wild Fauna and Flora
CMS	Convention on Migratory Species
CO_2	carbon dioxide
CO_2e	carbon dioxide equivalent
DOE	United States Department of Energy
EIA	environmental impact assessment
EIS	environmental impact statement
FAO	Food and Agriculture Organization of the United Nations
FSC	Forest Stewardship Council
GEO	Global Environment Outlook
GFCS	Global Framework for Climate Services
Gt	gigatonne
ha	hectare
HCFC	hydrochlorofluorocarbon
HFC	hydrofluorocarbon
HLW	high level waste
IAEA	International Atomic Energy Agency
ICRP	International Commission on Radiological Protection
ILW	intermediate level waste
IPBES	Intergovernmental Panel on Biodiversity and Ecosystem Services
IPCC	Intergovernmental Panel on Climate Change
IRENA	International Renewable Energy Agency
IUCN	International Union for Conservation of Nature
LLW	low level waste
LULUCF	land use, land-use change and forestry
MARPOL	International Convention for the Prevention of Pollution from Ships
MDG	Millennium Development Goal
MEA	Multilateral Environmental Agreement
MRV	measuring, reporting and verifying
N_2O	nitrous oxide
NASA	United States National Aeronautics and Space Administration
NEA-OECD	Nuclear Energy Agency of the Organisation for Economic Co-operation and Development
NOAA	United States National Oceanic and Atmospheric Administration
ODP	ozone depletion potential
ODS	ozone-depleting substance
OECD	Organisation for Economic Co-operation and Development
PEFC	Programme for Endorsement of Forest Certification
ppm	parts per million
REDD	Reducing Emissions from Deforestation and forest Degradation
RLI	Red List Index
SEPA	Scottish Environment Protection Agency
SOC	soil organic carbon
SOM	soil organic matter
UNCCD	United Nations Convention to Combat Desertification
UNCED	United Nations Conference on Environment and Development
UNCLOS	United Nations Convention on the Law of the Sea
UNCSD	United Nations Conference on Sustainable Development
UNDP	United Nations Development Programme
UNEP	United Nations Environment Programme
UNESCO	United Nations Educational, Scientific and Cultural Organization
UNFCCC	United Nations Framework Convention on Climate Change
UNSD	United Nations Statistics Division
VLLW	very low level waste
WMO	World Meteorological Organization

Executive Summary

In 2011, scientists made further progress in understanding our global environment. As countries look forward to the United Nations Conference on Sustainable Development (Rio+20) in Brazil in June 2012, evidence of certain environmental trends continues to accumulate, including those related to climate change and its impacts, biodiversity loss, and degradation of land and soils. Enormous challenges remain with respect to addressing the underlying causes and impacts of such trends, although positive developments can be reported as well, for instance in the area of renewable energy technology uptake and investment. Key environmental indicators such as those presented in the UNEP Year Book help keep track of the state of the environment by providing a picture of the latest available data and trends.

A feature of each Year Book is a review of environmental events and developments during the past year. In addition, each Year Book includes chapters that examine emerging environmental issues, written by groups of scientists who are experts in the field. The UNEP Year Book 2012 focuses on the vital role of soil carbon and the critical need to maintain and enhance it, in order to sustain its multiple economic, societal and environmental benefits. It also brings to the forefront some of the complexities and implications of the expected rapid increase in the number of nuclear reactors to be decommissioned in the next ten years.

2011 was a record-breaking year for extreme climate and weather events. Leading scientists are investigating the relationship between such events and climate change. According to the latest insights, climate change is leading to changes in the frequency, intensity, length, timing and spatial coverage of extreme weather events. New studies also suggest that the combined impacts of higher sea temperatures, ocean acidification, lack of oxygen and other factors could lead to the collapse of coral reefs and the spread of ocean dead zones. An increase in the total coverage of marine protected areas could halt some of the damage, provided these areas are established rapidly enough and are managed effectively, with the guidance of sound science.

In the face of further land use change and land use intensification to meet global demands for food, water and energy, sustaining or even enhancing soil carbon stocks becomes a priority. During the past 25 years, one-quarter of the global land area has suffered a decline in productivity and in the ability to provide ecosystem services due to soil carbon losses. Because soil carbon is central to agricultural productivity, climate stabilization and other vital ecosystem services, creating policy incentives around the sustainable management of soil carbon could deliver numerous short- and long-term benefits. In some locations, mechanisms will be needed to protect soils that are important soil carbon stores, such as peatlands and tundra, as alternatives to other uses such as agricultural or forestry expansion. However, in many cases multiple economic, societal and environmental benefits can be obtained on the same land through effective management of soil carbon.

A new focus at all levels of governance on effectively managing soil carbon for multiple benefits would constitute a significant step towards meeting the need for ecosystem services to support the world population in 2030 and beyond.

Nuclear decommissioning refers to safe handling, at the end of life, of nuclear power reactors and nuclear facilities. As the first generations of such reactors reach the end of their original design lives and some countries review their nuclear power programmes in the wake of the Fukushima accident, the number of reactors to be decommissioned in the next ten years is set to increase significantly. Each decommissioning presents particular technical challenges and risks to human health and the environment. Although decommissioning has been carried out for a number of years without major radiological mishaps, there are considerable geographical differences in expertise. The cost of decommissioning varies greatly, depending on the reactor type and size, its location, the proximity and availability of waste disposal facilities, the intended future use of the site, and the condition of both the reactor and the site at the time of decommissioning. It represents a substantial share of the cost of a nuclear power reactor's overall operations.

Decommissioning typically generates two-thirds of all the very low, low, and intermediate level waste produced during a reactor's lifetime. As the number of nuclear power plants scheduled for decommissioning grows, countries need to be prepared to handle these levels of waste. The scale of the task ahead will require national and international regulation, extensive funding, innovative technology and large numbers of trained workers. One lesson that begins to emerge is that nuclear power plants should be designed from the start for safe and efficient decommissioning.

Year in Review
Environmental events and developments

2011 was a year of environmental extremes. Major droughts and flooding were prominent in the news, and leading climate scientists continued their work to establish whether there is a clear relationship between extreme weather events and climate change. In the ocean, as few as 9 per cent of all species may have been identified, yet new studies show that overfishing, pollution and climate change severely threaten the future of ocean life. Despite the economic recession, global investments in green energy grew by nearly a third to US$211 billion in 2010. An investment of 2 per cent of GDP in ten key sectors could significantly accelerate the transition to a more sustainable, low-carbon economy.

Some 13 million people in Djibouti, Eritrea, Ethiopia, Kenya and Somalia have been experiencing one of the worst humanitarian crises in decades. The region's most severe drought in 60 years has caused widespread starvation and made access to clean water and sanitation extremely difficult **(Box 1)**. These conditions not only directly affect local communities today, but also weaken their resiliency to cope with future droughts, diminishing prospects for water and food security in the years to come (Munang and Nkem 2011). Temperatures in the region are expected to continue rising while rainfall patterns change (Anyah and Qui 2011).

The crisis in the Horn of Africa is only one of the events in 2011 that exemplify the challenges to be met in the face of an increasingly variable and changing climate worldwide. Many regions need innovative strategies to address pressures on land and water resources and on agricultural productivity – from building resilience in small-scale farming communities to global commitments to mitigate climate change.

Climate change, extreme weather events and disaster risk management

2011 was a year of record-breaking weather events, which caused a large number of deaths and billions of dollars in damage **(Figure 1)**. It was also the tenth warmest year and the warmest La Niña year on record, as well as the year in which the

Hydro-thermal vents are geysers on the seafloor supporting unique communities. Trawling and mineral mining can cause serious damage to deep sea ecosystems. *Credit: Charles Fisher*

Box 1: Drought response in the Horn of Africa

In 2011 the Dadaab refugee camp in Kenya became the home of 400 000 people fleeing drought and famine. *Credit: Linda Ogwell, Oxfam*

Drought, accompanied by high food prices, insufficient humanitarian action and restrictions on aid acceptance, has induced mass migration to refugee camps in the Horn of Africa. Famine warnings were issued for this region at the beginning of 2011, but the drought still had an extreme impact. In July the rate of acute malnutrition in southern Somalia had gone up to 38-50 per cent (FEWSNET 2011). Many early warning systems assess conditions on a country-by-country basis. Their ability to see the larger regional picture is therefore limited, which can affect the adequacy of response efforts (Ververs 2012).

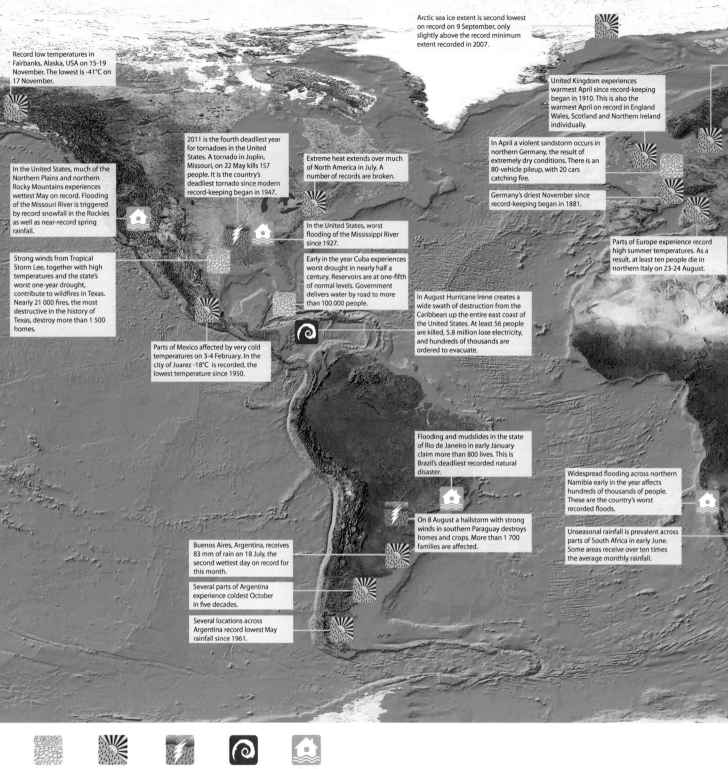

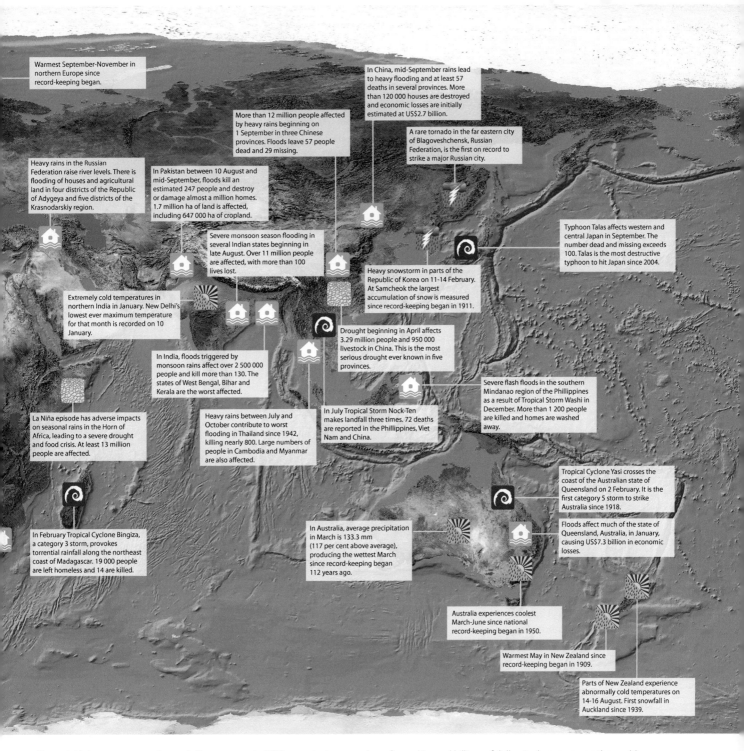

Figure 1: Major extreme weather and climate events in 2011 caused a large number of casualties and billions of dollars in damage across the world. Record-breaking temperatures and precipitation, as well as intense storms, tropical cyclones, floods, droughts and wildfires, resulted in many deaths and widespread destruction. According to the IPCC, climate change contributes to changing patterns in the frequency and intensity of extreme weather events (IPCC 2011a). A connection between climate change and geophysical events, such as earthquakes, has not been established.

second lowest seasonal minimum extent of Arctic sea ice was recorded (NSIDC 2011, WMO 2011a). Scientists have established a new international partnership to assess, on a case-by-case basis, the likelihood that exceptional weather events are caused or exacerbated by the global temperature increases observed during the past century (Stott et al. 2011). In addition, by investigating rainfall variability, scientists have already found evidence that anthropogenic greenhouse gas emissions substantially increase the risk of extreme events occurring (Pall et al. 2011).

An extreme weather event is defined as one that is rare within its statistical reference distribution at a particular location (IPCC 2011a). While natural variability makes it very difficult to attribute individual extreme weather events to climate change, statistical analyses show that the overall trends of many extreme events are changing. A new report of the Intergovernmental Panel on Climate Change (IPCC) concludes that climate change is leading to changes in the frequency, intensity, length, timing and spatial coverage of extreme events (IPCC 2011a). According to this report, it is virtually certain (99-100 per cent probability) that the frequency and magnitude of daily high temperatures will increase during the 21st century while those of cold extremes will decrease. The IPCC report expresses great confidence that there will be increases in events related to heavy precipitation and coastal high water, the latter due to rising sea levels. But despite a number of devastating floods in 2011, such as those in Australia, Pakistan and Thailand, evidence concerning regional long-term changes in flood magnitude and frequency is not as prevalent, partly because of a lack of available observational data at the appropriate time and spatial scales (IPCC 2011a).

The United States National Oceanic and Atmospheric Administration (NOAA) reported that in the first six months, 2011 had broken the record for the costliest year in terms of weather disasters in the United States (NOAA 2011). By the end of 2011, the United States experienced 14 "billion-dollar disasters" – disasters causing at least US$1 billion in damage (NOAA 2012). At the global level, in the first half of 2011 alone, costs arising from severe natural events exceeded those in the total previous costliest year, 2005 (UNISDR 2011). Munich Re, the world's largest reinsurance company, reported US$380 billion in losses in 2011 from natural disasters, which include weather and climate related events, as well as geophysical events such as earthquakes (Munich Re 2012). These staggering figures demonstrate the potential economic impact of an increase in frequency and severity of extreme weather events. They also suggest the degree of associated human suffering and the need for better risk reduction and preparedness strategies to increase resiliency to these events in both developed and developing countries.

Economic losses due to disasters are higher overall in developed countries than in developing ones. As a proportion of GDP, however, losses are much higher in developing countries. Over 95 per cent of extreme event fatalities in the past several decades have occurred in developing countries. Developed countries often have better financial and institutional mechanisms to cope with extreme events and their impacts. Future exposure and vulnerability to such events can be mitigated by integrating disaster risk reduction planning with economic development and climate change adaptation planning. Early warning and disaster risk reduction plans and strategies are essential, while documentation of individual events adds to the pool of knowledge and lessons learned (IPCC 2011a). Many regions are already carrying out disaster risk reduction and preparedness activities, including public awareness initiatives and improvements to early warning systems and infrastructure.

Population dynamics and resource scarcity

Extreme events can cause internal and external displacements of populations. In view of ongoing climate change and the likely increase in certain types of extreme events, the impact of these events on migration needs to be considered. More generally, there is the question of the implications of climate change for international security. In July 2011, the United Nations Security Council formally debated this issue, discussing ways in which climate change could be a "threat multiplier" in regard to maintaining global peace and stability. Environmental refugees displaced by water shortages and food crises are reshaping the world's human geography. While there was debate among the 15 Security Council members on the level of priority that should be associated with climate change, a statement was agreed

Secretary-General Ban Ki-moon (centre) with students from the New Explorations into Science, Mathematics and Technology School holding "7 000 000 000" signs the week the world population reached 7 billion. *Credit: Eskinder Debebe*

which expressed "concern that possible adverse effects of climate change may, in the long run, aggravate certain existing threats to international peace and security" (UN Security Council 2011).

A study published in December by four UN agencies looking at climate change, migration and conflict in the Sahel region of West Africa found that the Sahel is already experiencing changes in climate trends (UNEP 2011a). These changes are having an impact on the availability of natural resources and on food security, and are leading to shifts in migration patterns. The study looked at increased competition for natural resources, mainly land and water, resulting in conflicts among different communities and livelihood groups. In Darfur in East Africa, migration patterns are also putting a great strain on natural resources, including water. Half of Darfur's population now lives in and around urban areas. Before the civil conflict, only 20 per cent of the population was urban (UNHCR 2010). This unplanned urbanization has led to informal settlements with poor sanitation and waste management.

In 2011 the world population reached 7 billion. It is expected to grow to 9 billion by 2043, placing high demands on the Earth's resources (UN DESA 2011) **(Figure 2)**. Climate change exacerbates pressures to meet a growing and wealthier population's need for food. Global agricultural production may have to increase 70 per cent by 2050 to cope with this demand (FAO 2011a). A recent analysis of historical data shows that observed climate trends have had negative impacts on wheat and maize yields in the past 30 years (Lobell et al. 2011). Resource consumption could triple by 2050, while current consumption trends differ greatly between developed and developing countries (UNEP 2011b). For many agricultural systems there is the danger of a progressive breakdown of productive capacity under a combination of excessive population pressure and unsustainable agriculture use and practices (FAO 2011b).

Climate change, which will affect rainfall patterns in many regions, is expected to exacerbate water scarcity. This is of particular concern in high-intensity food-producing regions. Farming methods that are more environmentally sound need to be used, such as improved irrigation techniques and planting of vegetative cover including trees and shrubs to reduce water runoff and increase protection against drought (UNEP and IWMI 2011). Bioenergy production can put increased stress on land and water, competing with the need to feed the world's increasing population. The use of biofuels instead of fossil fuels, however, can also help reduce greenhouse gas emissions. Sustainability standards need therefore to be carefully defined

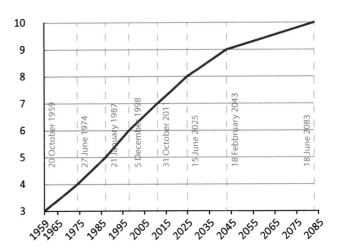

Figure 2: World population prospects in billions of people, 1959-2085. *Source: UN DESA (2011)*

and applied to ensure that rising demand for bioenergy does not lead to greater stress on land, water and food production (UNEP 2011c). Policies that protect both the land used for bioenergy production and surrounding ecosystems are necessary to maintain food and water security. Integrated planning and management can reduce the risks associated with the use of biofuels and still contribute to the building of a green economy (UNEP 2011c, UNEP et al. 2011).

In June 2011 governments attending the World Meteorological Organization (WMO) congress endorsed the Global Framework for Climate Services (GFCS), a co-ordinated effort by many stakeholders to make climate information for decision making and adaptation more accessible. The goal of the GFCS is to mainstream climate information for use across all countries and climate-sensitive sectors. Good co-ordination with climate financing activities and several tens of millions of dollars would be necessary to kick-start implementation of the GFCS to better support developing countries (WMO 2011b). One initiative that supports the GFCS is the Programme of Research on Climate Change Vulnerability, Impacts and Adaptation, launched in 2011 (PROVIA 2011). Developing countries have repeatedly asked for more co-ordinated science development to help national and sectoral adaptation strategies, plans and programmes. This initiative has the potential to meet some of these demands.

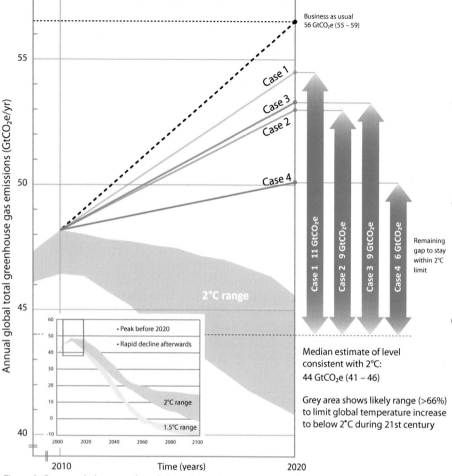

Figure 3: Country pledges to reduce their emissions by 2020 are currently not adequate to stay below a target global temperature rise of 2°C by the end of the 21st century, resulting in a gap. The size of the gap depends on the extent of pledges that are implemented and how they are applied. Four cases are considered in the figure: Case 1 reflects lower-ambition reduction pledges by countries and "lenient" accounting rules; Case 2 reflects lower-ambition reduction pledges and "strict" accounting rules; Case 3 represents more ambitious reduction pledges and "lenient" accounting rules; and Case 4 reflects more ambitious reduction pledges and "strict" accounting rules. Under "lenient" rules, allowances from land use, land-use change and forestry (LULUCF) accounting and surplus emissions credits can be counted towards a country's emissions pledges. Under "strict" rules, they cannot be used. Source: UNEP (2011d)

Energy and climate change mitigation

While many countries are taking steps to adapt to climate change, curbing global greenhouse gas emissions remains crucial to avoid the most severe and irreversible climate change impacts. In 2010, greenhouse gas levels were the highest recorded since preindustrial times (WMO 2011c). Many countries made pledges in 2009 to reduce their emissions of greenhouse gases by 2020, with the aim of keeping global warming below 2°C by the end of the 21st century. However, a significant gap of 6-11 Gt of CO_2 equivalent remains between expected levels of emissions in 2020 (based on current trends) and levels consistent with keeping the increase in the global average temperature by the end of the century from exceeding 2°C (**Figure 3**).

Cutting emissions by 2020 in a way that would limit the temperature rise to 2°C or less is economically and technologically feasible (UNEP 2011d). To cut emissions, countries need to shift their energy systems by increasing the use of existing low-carbon renewable energy sources and improving energy efficiency. Sector-specific policies to reduce emissions can be implemented, especially policies related to electricity production, industry, transport, forestry and agriculture. Such actions can help close the gap between current emission levels and emission

> **Box 2:** The Durban climate change negotiations
>
> The 17th Conference of the Parties to the UNFCCC (COP17) and the 7th Session of the Conference of the Parties serving as the Meeting of the Parties to the Kyoto Protocol (CMP7) were held in Durban, South Africa, on 28 November–9 December 2011. At stake was the need to reach a decision on a successor to the Kyoto Protocol (adopted in 1997), under which developed countries are committed to quantified emission reductions, as the first commitment period of the Protocol was scheduled to end in 2012.
>
> After prolonged debate, the Kyoto Protocol was extended into a second commitment period. Arrangements are to be finalized by the end of 2012 for its entry into force from 1 January 2013. Without several big emitters and with a bottom-up approach to setting emission reduction commitments, the second period of Kyoto may only serve as a transition to the universal and comprehensive agreement. Shortly after the Durban climate talks, Canada announced its withdrawal from the Kyoto Protocol.
>
> Complementary to the extension of the Kyoto Protocol, there was a landmark decision to start negotiating a protocol or a legal instrument or an agreed outcome with legal force under the Convention under the new track of the Durban Platform, which would include both developed and developing countries. Emission reductions under the new global agreement should start in 2020. Another important part of the agreements in Durban was the operationalization of the Green Climate Fund. Broad agreement was reached on the structure of this Fund. There was also a reiteration of the earlier goal of mobilizing jointly US$100 billion per year by 2020 to address the needs of developing countries. With the operationalization of the Green Climate Fund, climate finance may become more centralized and coherent.
>
> While some progress was made in Durban in ensuring that climate negotiations remain on track, there is concern that not enough progress was made in addressing the emissions gap. Current voluntary emission reduction pledges have so far not resulted in a reduction of global greenhouse gas emissions. Instead, these emissions have increased. The next meeting of COP18/CMP8 is planned for 26 November-7 December 2012 in Doha, Qatar.
>
>
> Credit: Siemens AG

targets, along with more ambitious reduction pledges and stricter accounting. Green procurement can also contribute to emission reductions by the public and private sector. It can be practised by individual businesses and organizations. Green procurement involves selecting those services and products that minimize impacts on the environment, including through greenhouse gas emission reductions. It results in organizations that are more environmentally responsible and often results in cost savings for these organizations as well (IISD 2011).

International negotiations under the United Nations Framework Convention on Climate Change (UNFCCC) are playing an important role in setting greenhouse gas emission reduction targets for countries. The Durban climate talks resulted in agreements on a second commitment period for the Kyoto Protocol and a process to start negotiating a legal instrument or an agreed outcome with legal force under the Convention covering all countries (**Box 2**). In many ways these two agreements symbolize a breakthrough. In addition, the Durban decisions operationalized the Green Climate Fund and furthered the established Cancun architecture for climate change, including a process to establish details of the Climate Technology Centre and Networks. However, the Durban decisions did not help to put in place a process for reducing emissions in line with what the science says is required to keep temperature increase below 2°C. There still remains a major emissions gap.

Limiting emissions of hydrofluorocarbons (HFCs) can make an important contribution to reducing total greenhouse gas emissions to prevent dangerous climate change (UNEP 2011e). Although HFCs are potent greenhouse gases, they have been used increasingly as substitutes for ozone-depleting substances such as chlorofluorocarbons (CFCs) and hydrochloroflurocarbons (HCFCs). The contribution of HFCs to total climate forcing is less than 1 per cent of that of all other greenhouse gases combined, but between 2004 and 2008 their use increased by about 8 per cent per year. The increase in HFC emissions could therefore have a noticeable impact on the climate system. HFC use can be reduced through the implementation of technical options, such as the substitution of architectural designs that avoid the need for air conditioning and the use of low global warming potential HFCs, which scientists are currently developing and introducing (UNEP 2011e).

Multiple immediate benefits can be obtained by reducing emissions of black carbon and chemicals that are precursors to ground-level ozone formation (Shindell et al. 2012). Black carbon is particulate matter formed through incomplete combustion of biomass and fossil fuels. Tropospheric ozone is a secondary pollutant, produced by chemical reactions of certain compounds in the presence of sunlight. One of the main precursors of tropospheric ozone is methane, which is also a powerful greenhouse gas. Both tropospheric ozone and black carbon affect the climate system and have significant impacts on human and ecosystem health (UNEP and WMO 2011). They also affect rainfall patterns and regional circulation patterns, such as the Asian monsoon. Black carbon darkens snow and ice, reducing the amount of sunlight reflected back into space. This causes warming and increased snow melt and consequently flooding. Targeting emissions of black carbon and ozone precursors has immediate benefits for human health and could help to mitigate climate change in the near term (Shindell et al. 2012). Effective actions to reduce CO_2 emissions are, however, still required in order to remain within a 2°C temperature rise.

A global view of black carbon on 26 September 2009, using data from NASA's GEOS5 GOCART climate model. Although challenges in measuring the global distribution of black carbon remain, scientists are using satellite data and computer models to better understand how black carbon particles influence the Earth's clouds and climate on a short-term basis. Aerosol optical thickness ranges non-linearly from 0.002 (transparent) to 0.02 (purple) to 0.2 (white). Source: NASA (2010)

black carbon aerosol optical thickness
0.002 0.005 0.01 0.02 0.05 0.1 0.2

Several approaches, such as improved energy conservation and efficiency, can be used effectively in conjunction with renewable energy technologies to reduce total greenhouse gas emissions. Some investment is required to obtain the maximum benefit of these approaches. According to a new study, an investment of 2 per cent of global GDP across ten key sectors is necessary to prompt a shift to a low-carbon, resource-efficient and socially inclusive green economy (**Box 3**). While job losses in some sectors would be inevitable, job creation in the longer term is expected to offset short-term losses. In 2011 several UN and other organizations jointly published guidelines for the transformation to a green economy (UN 2011a, UNEP 2011f).

Box 3: Ten key sectors for a green economy

An investment of 2 per cent of global GDP across ten key sectors is necessary to prompt a shift to a low-carbon, resource-efficient and socially inclusive green economy. These sectors are:

- Agriculture
- Fisheries
- Water
- Forests
- Renewable energy
- Manufacturing
- Waste
- Building
- Transport
- Tourism

A host of renewable energy solutions exist or have been proposed and are at various stages of development. Six categories of renewable energy technologies, in particular, have potential to mitigate climate change in the present or in the near future (IPCC 2011b) (**Box 4**). In 2008 renewable energy accounted for 12.9 per cent of total primary energy supply. Investment in renewable energy grew by 32 per cent between 2004 and 2008 to US$211 billion, with China emerging as a leader in the development of renewable energy technologies (REN21 2011, UNEP 2011f). Investment in renewable energy is projected to double to US$395 billion by 2020 (Bloomberg 2011). Renewable energy could account for 77 per cent of total primary energy supply by 2050 (IPCC 2011b).

In April 2011 the first session of the International Renewable Energy Agency (IRENA) assembly took place. This organization is focusing on the use of renewables as a tool for development, as well as on facilitating knowledge and technology transfer, adopting policies that promote renewable energy, and creating partnerships with relevant stakeholders to promote financing of renewable energy projects. As part of the UN Secretary-General's initiative to promote renewable energy, energy efficiency and universal access to modern sources of energy by 2030, the year 2012 has been declared the Year of Sustainable Energy for All (UN 2012).

Box 4: Renewable energy technologies to combat climate change

Hydropower projects need solid planning and management to avoid unintended environmental and social impacts. Credit: Hydro Pacific

- **Bioenergy** can be produced from agricultural, forestry and livestock residues, energy crops, and other organic waste streams. A wide range of these technologies exists, and they vary greatly in their technical maturity.
- **Direct solar energy** technologies harness the sun's energy to produce electricity and heat. Solar energy is variable and intermittent, producing different amounts of power on different days and at different times of the day. Relatively mature solar energy technologies exist.
- **Geothermal energy** is produced from the thermal energy in the Earth's interior. Geothermal power plants, which extract energy from reservoirs that are sufficiently permeable and hot, are fairly mature technologies. Geothermal energy can also be used directly for heating.
- **Hydropower** is produced by harnessing the energy of water that moves between different elevations. Hydropower technologies are very mature. Reservoirs often have multiple uses in addition to electricity production, such as support for drinking water availability, drought and flood control, and irrigation.
- **Ocean energy** harnesses the thermal, kinetic and chemical energy of seawater. Most ocean energy technologies are still in the research and development or pilot phases.
- **Wind energy** is produced from the kinetic energy of moving air, using large on- and offshore wind turbines. Onshore technologies are widely manufactured and used, and further development of offshore technologies is promising. Wind energy is variable, and in some locations unpredictable, but research indicates that many technical barriers can be overcome.

As the generation of nuclear energy does not produce emissions of greenhouse gases like the burning of fossil fuels, there has been increased interest in this type of energy in the past decade. The Fukushima nuclear power plant accident in March 2011, a series of equipment failures that followed a devastating 8.9 magnitude earthquake and tsunami, has further stimulated debate on nuclear energy's role in a secure and sustainable energy future. In 2010, 13.5 per cent of total global energy production came from nuclear power plants. At 74.1 per cent, France has the highest proportion of electricity generation from nuclear sources (NEI 2011).

Germany has announced plans to shut down all its nuclear power plants by 2022. Nuclear energy made up 27.3 per cent of its total electricity production in 2010 (NEI 2011). Germany plans to invest much more in renewable energy. France, on the other hand, has announced that it will invest US$1.4 billion in additional nuclear power development. This will include investments in research on safety. The closing and decommissioning of nuclear power reactors – an emerging international issue – is the topic of Chapter 3 of this *Year Book*.

Most human-induced greenhouse gas emissions derive from fossil fuels, which are still the world's main energy source. The expansion of oil exploration activities continues, particularly in the Arctic region. For instance, in 2011 the United States government announced that it would move forward with leases for exploration off the coast of Alaska. It released a five-year plan under which 75 per cent of estimated oil and gas resources would be made available for exploration off the Alaskan coast and in areas of the Gulf of Mexico (US DOI 2011). Oil exploration in the Arctic is increasing partly because melting sea ice is allowing oil tankers to expand their routes into previously inaccessible areas. Human activity is expected to continue to increase in the polar regions. Environmentalists have expressed concerns about this development, mainly related to possible oil spills **(Box 5)**.

Improvements in technologies for horizontal drilling and hydraulic fracturing have made it economically feasible to produce large volumes of natural gas, particularly shale gas, from low-permeability geological formations (a process known as "fracking"). Fracking typically involves high pressure injection of chemicals deep underground, blasting fractures in geological formations to release gas **(Figure 4)**. The most significant development and exploitation of shale gas and other unconventional types of natural gas has taken place in North America.

Despite the considerable economic benefits of producing and using shale gas and other types of unconventional gas (e.g. job

Box 5: Impact of oil spills

Oil contamination at the Bomu flow station in K-Dere, Ogoniland, Nigeria. *Credit: UNEP*

The expansion of oil drilling in the Arctic brings with it potential risks. A major well blowout is more likely during drilling of the first exploratory well of a geological structure than at any other time. Off-shore spill preparedness is not always in place to deal with such a risk (Porta and Bankes 2011). Specific standards are important to avoid the most negative impacts of oil spills. The 2010 spill in the Gulf of Mexico received widespread media attention and provoked public outcries. Oil spills in Nigeria have received far less international attention, although they have been at the heart of social unrest for decades. A study conducted at the invitation of the Nigerian government evaluates the environmental and health impacts of oil contamination in the country's Ogoniland region (UNEP 2011g). It concludes that widespread oil pollution in Ogoniland is severely impacting the environment and is posing serious health risks in some communities.

Hydrocarbon pollution has reached very high levels in soil and groundwater at a majority of the sites examined. Residents of Ogoniland have been exposed to chronic pollution as a result of oil spills and oil well fires, increasing cancer risks. Crops have been damaged and the fisheries sector has suffered due to persistent hydrocarbon contamination of many of the region's creeks. The study estimates that cleaning up the pollution and catalyzing a sustainable recovery of Ogoniland could take between 25 and 30 years. It therefore calls for emergency measures to minimize dangers to public health, and for long-term co-ordinated action to achieve environmental restoration.

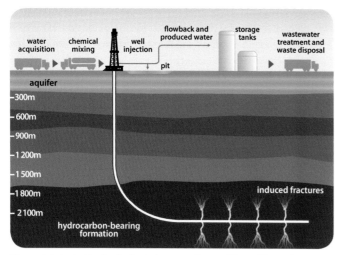

Figure 4: In typical hydraulic fracturing operations, millions of litres of water, chemicals and sand are injected at high pressure down a well. The pressurized fluid mixture causes the rock formation to crack, allowing natural gas or oil to flow up the well. *Source: Adapted from US EPA (2011)*

creation, greater energy independence), fracking is controversial because of widespread concerns about its health and environmental effects (Osborn et al. 2011, US EPA 2011, Cathles et al. 2012). These concerns include:

- drinking water contamination, which can result from the injection of chemicals deep underground during the fracking process;
- the greenhouse footprint of fracking operations, especially fugitive methane emissions; and
- seismic activity, which can occur when water or other fluids are injected deep underground during this process.

The United States Energy Information Agency has published assessments of 48 shale gas basins in 32 countries, containing almost 70 shale gas formations (US EIA 2011). While these assessments are likely to change as additional information becomes available, they show that the international shale gas resource base is potentially vast. As fracking spreads to new parts of the world, consideration needs to be given to its impact on health and the environment in countries where, among other differences, there is little experience with fracking operations.

Global biodiversity conservation

2011 was the International Year of Forests, during which a number of events were dedicated to their protection and sustainable development. Forests are of vital importance to biodiversity and the global economy. The livelihoods of 1.6 billion people depend on them (UN 2011b). Deforestation and forest degradation contribute 15-17 per cent of global greenhouse gas emissions (UN-REDD 2011). In 2010, the UNFCCC Cancun Agreement supported Reducing Emissions from Deforestation and forest Degradation (REDD+) in developing countries as a means of placing financial value on the carbon stored in forests. At the climate meeting in Durban further progress was made on the full mechanism, with safeguards and options for results-based financing for which market-based approaches could be developed.

Not only does vegetation on land, especially that in forests, absorb CO_2, but sea grass beds, mangroves, mudflats and other coastal wetlands also sequester it. However, the increasing human impacts on coastal areas, for instance from settlements and aquaculture, have destroyed an estimated 65 per cent of sea grass and wetland habitats (Lotze et al. 2006). Coral reefs are one of the world's most biodiverse ecosystems, providing a range of benefits to society. They supply resources for the development of new products by the international pharmaceuticals industry, provide habitat for a quarter of the world's fish biodiversity and support local economic development. Scientists warn that life in the ocean is being severely threatened by overfishing, pollution and climate change (Rogers and Laffoley 2011). For instance, one-third of fish in the Indian Ocean are at risk of local extinction (Graham 2011). The combined impacts of factors such as higher sea temperatures, ocean acidification and lack of oxygen may lead to the collapse of coral reefs and the spread of ocean dead zones (Rogers and Laffoley 2011). In August 2011 leading scientists associated with the Census of Marine Life project, a decade-long assessment of the world's oceans completed in 2010, presented their findings concerning human impacts on the deep seas (Ramirez-Llodra et al. 2011) **(Box 6)**.

Recent research indicates that only 14 per cent of the world's species are known (Mora et al. 2011). In the ocean, as few as 9 per cent of all species may have been identified. This lack of knowledge raises critical questions about how we can adequately conserve global biodiversity, especially in the face of climate change. Gaps in scientific knowledge can make it difficult to protect the deep sea environment. Moreover, an overarching legal framework for the protection of oceans is lacking. This gap has been identified as an emerging challenge for the 21st century through the UNEP Foresight Process (UNEP 2012).

Box 6: Human impacts on the deep seas

Bioluminescent creatures create their own light in deep sea environments. *Credit: Monterey Bay Aquarium Research Institute*

- Around 6.4 million tonnes of litter per year ends up in the ocean. Plastic in the ocean is of particular concern, as it persists and not enough is known about the effects of microplastics in the ocean environment. Concerns have been raised that chemicals transported by such particles may enter the food chain (UNEP 2011h).
- Deep-sea trawling and mining practices are damaging the habitats of species that are often long-lived and reproduce slowly, and hence not well equipped to respond to increasing pressures.
- The main concern for the future is climate change, as the ocean's increasing acidity affects the ability of corals and shellfish to make skeletons and shells.

One way to stop some of the damage to ecosystems is to create protected areas. Governments at the meeting of the Convention on Biological Diversity (CBD) in October 2010 set a goal of increasing the coverage of marine protected areas ten-fold, from 1 per cent to 10 per cent, by 2020 (CBD 2010). The target for terrestrial protected areas is to increase their extent to 17 per cent. However, the effectiveness and current rate of establishing new protected areas may not be sufficient to overcome current trends in biodiversity loss (Mora and Sale 2011). There are problems related to gaps in the coverage of critical areas and to management effectiveness where there is strong pressure to develop.

On land, poaching took a high toll on large mammals in 2011. The western black rhinoceros was officially declared extinct by the International Union for Conservation of Nature (IUCN) following decades of poaching (IUCN 2011). In South Africa, 448 rhinos were killed in 2011 – up from 13 in 2007. At the time of writing, the number of rhinos poached in 2012 in South Africa had already reached 28 (SA DoEA 2012). Globally, 2010 saw the highest levels of elephant poaching since 2002, with central Africa causing the greatest concern (CITES 2011a). Poverty, poor governance, and increasing demand for ivory continue to drive poaching activity. The value of the ivory from a large male elephant is equivalent to 15 years' salary for an unskilled worker (Wittemyer et al. 2011).

Despite international agreements on ivory trading and the progress made in some countries, domestic and international trade bans are not enforced to the extent necessary to protect species. 2011 was the worst year in decades, with a number of large ivory seizures. An estimated 23 tonnes, for which some 2 500 elephants had been killed, was impounded from the year's 13 largest ivory seizures (TRAFFIC 2011). For the most part, this ivory was reportedly destined for Asia.

Illegal trade involves fraudulent applications for CITES documents, abuse of legal trophy hunting, and the use of couriers to smuggle horns. The African Elephant Action Plan, launched in 2011, is expected to enhance law enforcement capacity to protect against elephant poaching and illegal ivory trade. The International Consortium on Combatting Wildlife Crime began a programme in 2011 under which perpetrators of serious wildlife crimes will face a strong co-ordinated response, in contrast to the current situation where the risk of detection and punishment is low (CITES 2011b).

In Malaysia more than 3 000 tusks were seized in a period of three months in 2011, demonstrating a marked improvement in enforcement in that country. The increase in poaching rates correlates with that in ivory prices. Credit: ©TRAFFIC Asia

The plummeting numbers of animals at the top of the food chain, such as wolves, lions and sharks, is one of humanity's most pervasive influences on the natural world (Estes et al. 2011). The loss of such "apex consumers", largely due to hunting and habitat fragmentation, triggers a complex cascade of changes in ecosystems. The extent to which this is reshaping ecosystems is undervalued, as such top-down effects are difficult for scientists to demonstrate. However, as changes in the environment occur more rapidly, the need to strengthen the interaction between science and policy in order to ensure that decision making is based on sound science is becoming even greater.

Climate change, considered a threat multiplier for biodiversity, could drive the mass migration of numerous plant and animal species in coming years. These changes could further threaten species survival, significantly impacting the Earth's energy, carbon, water and biogeochemical cycles. By the year 2100, 40 per cent of land areas such as grassland or tundra could shift to a different state (Bergengren et al. 2011). For the first time, scientists have developed a model that assesses how animals respond to climate change in terms of both behaviour and genetics (Coulson et al. 2011). The model was developed based on longitudinal data from studies of grey wolves in Yellowstone National Park in the United States. It is expected that this model can help predict the climate change responses of many groups of animals.

Looking ahead

Scientists warn that the environment has been changing quickly from a period of stable state in which civilization developed during the past 12 000 years (the Holocene) to an unknown future state with significantly different characteristics (which some refer to as the Anthropocene) (Steffen et al. 2011). With population growth, some of the short-term solutions of the past, such as migrating when the environment is badly damaged or becomes less productive, are no longer viable. As demonstrated in the Arctic and on the ocean floor, today the impacts of human activities are felt far beyond our immediate surroundings.

The Earth is a complex system with highly interlinked components, some of which (such as soils) are greatly undervalued. For instance, the multiple benefits of soil carbon, described in Chapter 2 of this *Year Book*, are just starting to receive attention outside the realm of soil scientists. Earth system science is still in its infancy, but some scientists claim that humanity has already gone beyond the boundaries for climate change, biodiversity loss and excess production of nutrients, notably nitrogen and phosphorus (Rockström et al. 2009). Other areas identified as most in need of limitation are stratospheric ozone depletion, ocean acidification, global consumption of freshwater, changes in land use for agriculture, and air and chemical pollution.

Thanks to new methods of communication and observation, our understanding of the complexity of environmental issues is increasing. Many of the decisions we make affect the ecosystems that form the life support system upon which we depend. Scenarios for the future can help us look ahead and weigh the impact of our choices. For instance, a study in the United Kingdom examined scenarios of urban growth patterns (Eigenbrod et al. 2011). Under a scenario of dense housing growth, urban areas would experience a reduction in their abilities to cope with floods – a service considered important in light of the predicted climate change impact of more frequent and intense extreme weather events. They would not experience this effect under a low-density housing scenario, but there would be a reduction in the amount of land available for food and for carbon storage in the soil, services important for feeding a growing population and mitigating climate change. With smart planning and informed decision making guided by science, there are opportunities to maximize the benefits under both scenarios.

Such trade-offs, and the cost of actions versus the cost of inaction, also need to be considered from an international viewpoint. At the global level, discussions on many such actions will take centre stage at the United Nations Conference on Sustainable Development (Rio + 20) in June 2012. There will be a focus on the institutional framework for sustainable development and on the development of a green economy in the context of sustainable development and poverty eradication.

Reviewing new science and developments during the past year, concerns about population growth, resource use, climate change, widespread pollution and biodiversity loss all call for actions from the local to global levels to respond to sustainable development challenges. One of the world's champions of on-the-ground action, Professor Wangari Maathai, Nobel Peace Prize Laureate and founder of Kenya's Green Belt Movement, sadly passed away in September 2011. There is a need to continue her environmental work. Local leaders, civil society, companies and policy makers worldwide have an important role to play in overcoming some of the greatest environmental sustainability challenges.

A crowded market in Dhaka, Bangladesh. While the world's population is rapidly growing, action to address environmental challenges is critical to meet growing demands for food and ensure sustainable development. *Credit: IFPRI*

2011
At a glance

January — World food prices reach a historic peak for the seventh consecutive month.

February 2 — The United Nations International Year of Forests begins with the launch of FAO's 2011 State of the World's Forests report, which emphasizes that the forest industry can play an important role in a greener economy.

March 11 — The Fukushima nuclear power plant in Japan experiences a series of equipment failures following a severe earthquake and tsunami.

March 20-25 — Participants in the Fifth International Marine Debris Conference agree to the Honolulu Commitment, which outlines several approaches to the reduction of marine debris and calls for public awareness campaigns.

March 22 — The Roundtable for Sustainable Biofuels launches a global certification system during the World Biofuels Markets Congress in Rotterdam, the Netherlands. It is expected that this system will advance the sustainability of the global biofuels industry.

May 8-13 — The Third Global Platform for Disaster Risk Reduction results in pledges to improve disaster preparedness.

May 14-15 — The UN backs World Migratory Bird Day with a focus this year on land use and land sustainability.

May 16 - June 3 — The WMO Congress, the organization's supreme body, meets in Geneva to discuss the WMO's strategic direction for 2012-2015.

July 20 — The UN officially declares famine in two regions of Somalia, the first time a famine has been declared by the UN in almost 30 years.

July 20 — The UN Security Council holds a special meeting to consider its role in addressing climate change. The Secretary-General warns of climate change threats to international peace and security.

August 1 — A new ban on pollution from heavy grade fuel oils goes into effect in the Antarctic region, through amendments to the International Convention for the Prevention of Pollution from Ships (MARPOL).

August 26 — Participants in World Water Week release the Stockholm Statement, calling for an increase in water use efficiency and availability of water for all.

Sept 7-9 — The Global Soil Partnership is launched at FAO. The partnership aims at fostering favourable policies that provide technical expertise for soil protection and management.

Sept 26 — Professor Wangari Lauraete Maathai, Nobel Peace Prize Laureate and founder of Kenya's Green Belt Movement, passes away in Nairobi at age 71.

Oct 1 — The UNEP Tunza International Children and Youth Conference concludes with the endorsement of the Bandung Declaration, which calls on participants in the Rio+20 meeting to consider the needs of children and youth.

Oct 3-7 — The Intergovernmental Panel on Biodiversity and Ecosystem Services (IPBES), the new UN biodiversity forum, holds its first session in Nairobi, Kenya.

Oct 10-21 — The UNCCD COP10 meeting takes place in the Republic of Korea, exploring ways to advance efforts on desertification, land degradation and drought.

Oct 21 — Representatives of the 118 members of the Basel Convention reach an agreement to unblock an amendment banning the export of hazardous wastes from OECD to non-OECD countries.

Oct 31 — The world's population reaches 7 billion, increasing concerns about how the world will provide food and water to its growing population in the future.

Nov 7 — The IUCN and the CBD secretariat sign an invasive species agreement that will work towards identifying invasive species and their pathways.

Nov 8 — UNEP announces that the Billion Tree Campaign has reached its 12 billion landmark. The campaign aims to improve the quality of life in communities through multiple benifits.

December 11 — The Durban Platform is adopted at the UNFCCC COP17/CMP7 in Durban, South Africa. The Platform extends the life of the Kyoto Protocol and establishes the structure of a Green Climate Fund.

December 15 — The United Arab Emirates Ministry of Environment and Water, the Environment Agency-Abu Dhabi, and UNEP sign the Eye on Earth Declaration in Abu Dhabi, which stresses the importance of sharing environmental data and using it for decision making.

2012 Calendar of events

January

- **January 12-13**: 2nd Session of the International Renewable Energy Agency (IRENA) Assembly in Abu Dhabi, United Arab Emirates (UAE)
- **January 16-19**: 5th World Future Energy Summit, Abu Dhabi, UAE
- **January 22-27**: Arctic Frontiers Conference: Energies of the High North, Tromsø, Norway
- **January 23-27**: Global Conference on Land-Ocean Connections/3rd Intergovernmental Review on the Implementation of the Global Programme of Action for the Protection of the Marine Environment, Manila, the Philippines
- **January 31 - February 3**: Forum of Environment Ministers of Latin America and the Caribbean, Quito, Ecuador

February

- **February 5-6**: Second Asia-Pacific Water Summit, Bangkok, Thailand
- **February 20-22**: 12th Special Session of the UNEP Governing Council/Global Ministerial Environment Forum, Nairobi, Kenya
- **February 27**: OECD Green Skills Forum, Paris, France

March

- **March 12-17**: 6th World Water Forum, Marseille, France
- **March 19-22**: International Atomic Energy Agency (IAEA) International Experts Meeting on Reactor and Spent Fuel Safety in the Light of the Accident at the Fukushima Daiichi Nuclear Power Plant, Vienna, Austria
- **March 26-27**: 3rd Intersessional Meeting of the UN Conference on Sustainable Development, New York, USA
- **March 26-29**: "Planet under Pressure" conference, London, UK

April

- **April 16-21**: 2nd Session of the Plenary Meeting on the Intergovernmental Platform on Biodiversity and Ecosystem Services, Panama City, Panama
- **April 22-25**: B4E Business for the Environment – Global Summit, Berlin, Germany
- **April 22-27**: International Polar Year 2012 conference: From Knowledge to Action, Montreal, Canada

May

- **May 8-9**: 26th Session of North American Forest Commission, Quebec City, Canada
- **May 12-August 27**: Expo 2012 World's Fair with focus on "The Living Ocean and Coast", Yeosu, Republic of Korea
- **May 21-23**: Global Conference on Oceans, Climate and Security, Boston, USA
- **May 29-31**: 2nd International Climate Change Adaptation Conference, Tucson, USA

June

- **June 5**: World Environment Day - "Green Economy: Does it include you?"
- **June 13-15**: 3rd PrepCom for the UN Conference on Sustainable Development, Rio de Janeiro, Brazil
- **June 20-22**: United Nations Conference on Sustainable Development (Rio +20), Rio de Janeiro, Brazil

July

- **July 6-13**: 11th Meeting of the Contracting Parties to the Ramsar Convention on Wetlands of International Importance (COP11), Bucharest, Romania
- **July 9-13**: 30th Session of the FAO Committee on Fisheries, Rome, Italy
- **July 23-27**: 62nd Meeting of the CITES Standing Committee, Geneva, Switzerland
- **July 27-29**: From Science to Policy conference marking the 40th anniversary of the International Institute for Applied Systems Analysis, Vienna, Austria

August

- **August 6-9**: Pacific Rim Energy and Sustainability Congress, Hiroshima, Japan
- **August 29-31**: International Sustainability Conference, Basel, Switzerland

September

- **Sept 6-15**: International Union for Conservation of Nature World Conservation Congress 2012, Jeju, Republic of Korea
- **Sept 17-21**: 14th Session of the African Ministerial Conference on the Environment, Dar es Salaam, Tanzania
- **Sept 6-20**: Joint FAO/WHO Meeting on Pesticide Residues, Rome, Italy
- **Sept 17-21**: 3rd Session of the International Conference on Chemicals Management, Nairobi, Kenya
- **Sept 24-27**: Third International Symposium on the Ocean in a High-CO_2 World, Monterey, USA
- **Sept 24-26**: UNEP/GEF International Waters Science Conference, Bangkok, Thailand

October

- **Oct 8-19**: 11th Meeting of the Conference of the Parties to the UN Convention on Biological Diversity, Hyderabad, India

November

- **Nov 26-Dec 7**: 18th Session of the Conference of the Parties to the UN Framework Convention on Climate Change (UNFCCC) and 8th Session of the Meeting of the Parties to the Kyoto Protocol (COP18/CMP8), Doha, Qatar

December

- **Dec 11-13**: 4th International Conference on Sustainable Irrigation and Drainage: Management, Technologies and Policies, Adelaide, Australia

References

Anyah, R. and Qui, W. (2011). Characteristic 20th and 21st century precipitation and temperature patterns and changes over the Greater Horn of Africa. *International Journal of Climatology*. Published online 4 January. http://onlinelibrary.wiley.com/doi/10.1002/joc.2270/abstract

Bergengren, J.C., Walier, D.E. and Yung, Y.L. (2011). Ecological sensitivity: a biospheric view of climate change. *Climatic Change*, 107(3-4): 433-457

Bloomberg (Bloomberg New Energy Finance) (2011). Global Renewable Energy Market Outlook: Executive Summary. http://www.bnef.com/WhitePapers/download/53

Cathles, III, L.M., Brown, L., Taam, M. and Hunter, A. (2012). A commentary on "The greenhouse-gas footprint of natural gas in shale formations" by R.W. Howarth, R. Santoro, and Anthony Ingraffea. *Climatic Change*. Published online 3 January. http://www.springerlink.com/content/x001g12t2332462p/

CBD (Convention on Biological Diversity) (2010). A new era of living in harmony with nature is born at the Nagoya Biodiversity Summit. Secretariat of the Biological Diversity Convention, Montreal

CITES (2011a). Status of elephant populations, levels of illegal killing and the trade in ivory: a report to the standing committee of CITES. SC61 Doc. 44.2 (Rev.1) Annex 1. http://www.cites.org/eng/com/sc/61/E61-44-02-A1.pdf

CITES (2011b). The International Consortium on Combatting Wildlife Crime. http://www.cites.org/eng/prog/iccwc.shtml

Coulson, T., MacNulty, D.R., Stahler, D.R., von Holdt, B., Wayne, R.K. and Smith, D.W. (2011). Modeling Effects of Environmental Change on Wolf Population Dynamics, Trait Evolution, and Life History. *Science*, 334(6060), 1275-1278

Eigenbrod, F., Bell, V.A., Davies, H.N., Heinemeyer, A., Armsworth, P.R. and Gaston, K.J. (2011). The impact of projected increases in urbanization on ecosystem services. *Proceedings of the Royal Society*, 278(1722), 3201-3208

Estes, J.A, Terborh, J., Brashares, J.S., Power, M.E., Berger, J., Bond, W.J., Carpenter, S.R., Essington, T.E., Holt, R.D., Jackson, J.B.C., Marquis, R.J., Oksanen, L., Oksanen, T., Paine, R.T., Pikitch, E.K., Ripple, W.J., Sandin, S.A., Marten, S., Schoener, T.W., Shurin, J.B., Sinclair, A.R.E., Soulé, M.E., Virtanen, R. and Wardle, D.A. (2011). Trophic downgrading of planet earth. *Science*, 333(6040), 301-306

FAO (Food and Agriculture Organization of the United Nations) (2011a). *Climate change, water and food security*. http://www.fao.org/docrep/014/i2096e/i2096e00.pdf

FAO (2011b). *The State of the World's Land and Water Resources for Food and Agriculture: Managing Systems at Risk, Summary Report*. http://www.fao.org/nr/water/docs/SOLAW_EX_SUMM_WEB_EN.pdf

FEWSNET (Famine Early Warning System Network) (2011). Famine thresholds surpassed in three new areas of Southern Somalia, 3 August. http://www.fews.net/docs/Publications/FSNAU_FEWSNET_030811press%20release_final.pdf

Graham, N.A.J., Chabanet, P., Evans, R.D., Jennings, S., Letourneur, Y., MacNeil, M.A., McClanahan, T.R., Öhman, M.C., Polunin, N.V.C. and Wilson, S.K. (2011). Extinction vulnerability of coral reef fishes. *Ecology Letters*, 14(4), 341-348

IISD (International Institute for Sustainable Development) (2011). Green Procurement. http://www.iisd.org/business/tools/bt_green_pro.aspx

IPCC (Intergovenmental Panel on Climate Change) (2011a). Summary for Policymakers. In *Intergovernmental Panel on Climate Change Special Report on Managing the Risks of Extreme Events and Disasters to Advance Climate Change Adaptation*. Field, C.B., Barros, V., Stocker, T.F., Qin, D., Dokken, D., Ebi, K.L., Mastrandrea, M.D., Mach, K.J., Plattner, G.-K., Allen, S.K., Tignor, M. and Midgley, P.M. (eds.). http://www.ipcc-wg2.gov/SREX/images/uploads/SREX-SPM_Approved-HiRes_opt.pdf

IPCC (2011b). *Special Report on Renewable Energy Sources and Climate Change Mitigation*. Prepared by Working Group III of the Intergovernmental Panel on Climate Change. Edenhofer, O., Pichs-Madruga, R., Sokona, Y., Seyboth, K., Matschoss, P., Kadner, S., Zwickel, T., Eickemeier, P., Hansen, G., Schlömer, S. and von Stechow, C. (eds). http://srren.ipcc-wg3.de/report/IPCC_SRREN_Full_Report.pdf

IUCN (International Union for the Conservation of Nature) (2011). Red List Update: Another leap towards the barometer of life. Press release, 10 November. http://www.iucn.org/knowledge/news/?uNewsID=8548

Lobell, D., Schlenker, W. and Costa-Roberts, J. (2011). Climate Trends and Global Crop Production since 1980. *Science*, 333(6042), 616-620

Lotze, H.K., Lenihan, H.S., Bourque, B.J., Bradbury, R.H., Cooke, R.G., Kay, M.C., Kidwell, S.M., Kirby, M.X., Peterson, C.H. and Jackson, J.B.C. (2006). Depletion, Degradation, and Recovery Potential of Estuaries and Coastal Seas. *Science*, 312(5781), 1806-1809

Mora, C. and Sale, P.F. (2011). Ongoing global biodiversity loss and the need to move beyond protected areas: a review of the technical and practical shortcomings of protected areas on land and sea. *Marine Ecology Progress Series*, 434, 251-266

Mora, C., Tittensor, D.P., Adl, S., Simpson, A.G.B. and Worm, B. (2011). How Many Species Are There on Earth and in the Ocean? *PLoS Biology*, 9(8): e1001127. http://www.plosbiology.org/article/info:doi/10.1371/journal.pbio.1001127

Munang, R. and Nkem, J.N. (2011). Using small-scale adaptation actions to address the food crisis in the Horn of Africa: Going beyond food aid and cash transfers. *Sustainability*, 3(9), 1510-1516

Munich Re (2012). Global natural catastrophe update. 2011 natural catastrophe year in review. http://www.munichreamerica.com/webinars/2012_01_natcatreview/munichre_iii_2011natcatreview.pdf

NASA (2010). Global Transport of Black Carbon, Goddard Space Flight Center Scientific Visualization Studio. http://svs.gsfc.nasa.gov/vis/a000000/a003600/a003665/index.html

NEI (Nuclear Energy Institute) (2011). Nuclear Energy Around the World. http://www.nei.org/resourcesandstats/nuclear_statistics/worldstatistics/

NOAA (United States National Oceanic and Atmospheric Administration) (2011). Billion Dollar U.S. Weather/Climate Disasters. http://www.ncdc.noaa.gov/oa/reports/billionz.html

NOAA (2012). Extreme Weather 2011. http://www.noaa.gov/extreme2011/

NSIDC (United States National Sea and Ice Data Center) (2011). Arctic sea ice near record lows. http://nsidc.org/arcticseaicenews/2011/09/arctic-sea-ice-near-record-lows/

Osborn, S.G., Avner, V., Warner, N.R. and Jackson, R.B. (2011). Methane contamination of drinking water accompanying gas-well drilling and hydraulic fracturing. *Proceedings of the National Academy of Sciences of the United States*, 108(20), 8172-8176. http://www.pnas.org/content/108/20/8172.full

Pall, P., Aina, T., Stone, D., Stott, P., Nozawa, T., Hilberts, A., Lohann, D. and Allen, M. (2011). Anthropogenic greenhouse gas contribution to flood risk in England and Wales in autumn 2000. *Nature*, 470, 382-385

Porta, L. and Bankes, N. (2011). Becoming Arctic Ready: Policy Recommendations for Reforming Canada's Approach to Licensing and Regulating Offshore Oil and Gas in the Arctic. The Pew Environment Group. http://www.pewenvironment.org/uploadedFiles/PEG/Publications/Report/PewOilGasReport_web.pdf

PROVIA (Programme of Research on Climate Change Vulnerability, Impacts and Adaptation) (2011). What is Provia? http://www.provia-climatechange.org/ABOUT/WhatisPROVIA/tabid/55216/Default.aspx

Ramirez-Llodra, E., Tyler, P.A., Baker, M.C., Bergstad, O.A., Clark, M.R., Escobar, E., Levin, L.A., Menot, L., Rowden, A.A., Smith, C.R. and Van Dover, C.L. (2011). Man and the Last Great Wilderness: Human Impact on the Deep Sea. PLOS ONE, 6(8). http://www.plosone.org/article/info%3Adoi%2F10.1371%2Fjournal.pone.0022588

REN21 (Renewable Energy Policy Network for the 21st Century) (2011). *Renewables 2011 Global Status Report*. http://www.ren21.net/Portals/97/documents/GSR/REN21_GSR2011.pdf

Rockström, J., Steffen, W., Noone, K., Persson, Å., Chapin, III, F.S., Lambin, E.F., Lenton, T.M., Scheffer, M., Folke, C., Schellnhuber, H.J., Nykvist, B., de Wit, C.A., Hughes, T., van der Leeuw, S., Rodhe, H., Sverker Sörlin, S., Snyder, P.K., Costanza, R., Svedin, U., Falkenmark, M., Karlberg, L., Corell, R.W., Fabry, V.J., Hansen, J., Walker, B., Liverman, D., Richardson, K., Crutzen, P. and Foley, J.A. (2009). A safe operating space for humanity. *Nature*, 461, 472-475

Rogers, A.D. and Laffoley, D.d'A. (2011). *International Earth system expert workshop on ocean stresses and impacts. Summary workshop report*. IPSO (International Programme on the State of the Ocean), Oxford, UK. http://www.stateoftheocean.org/pdfs/1906_IPSO-LONG.pdf

SA DoEA (South African Department of Environmental Affairs) (2012). http://www.environment.gov.za/

Shindell, D., Kuylenstierna, J.C.I., Vignati, E., van Dingenen, R., Amann, M., Klimont, Z., Anenberg, S.C., Muller, N., Janssens-Maenhout, G., Raes, F., Schwartz, J., Faluvegi, G., Pozzoli, L., Kupiainen, K., Höglund-Isaksson, L., Emberson, L., Streets, D., Ramanathan, V., Hicks, K., Kim Oanh, N.T., Milly, G., Williams, M., Demkine, V. and

Fowler, D. (2012). Simultaneously Mitigating Near-Term Climate Change and Improving Human Health and Food Security. *Science*, 335(6065), 183-189

Steffen, W., Rockström, J. and Costanza, R. (2011). How defining planetary boundaries can transform our approach to growth. *Solutions*, 2(3). http://www.thesolutionsjournal.com/node/935

Stott, P.A., Allen, M., Christidis, N., Dole, D., Hoerling, M., Huntingford, C., Pall, P., Perlwitz, J. and Stone, D. (2011). Attribution of Weather and Climate-Related Extreme Events, WCRP OSC Climate Research in Service to Society, Denver, Colorado, USA. http://www.wcrp-climate.org/conference2011/documents/Stott.pdf

TRAFFIC (2011). http://www.traffic.org/home/2011/12/29/2011-annus-horribilis-for-african-elephants-says-traffic.html

UN (United Nations) (2011a). *Working towards a Balanced and Inclusive Green Economy: A United Nations System-wide Perspective. Prepared by the Environment Management Group*. http://www.unemg.org/Portals/27/Documents/IMG/GreenEconomy/report/GreenEconomy-Full.pdf

UN (2011b). Forests for People: Fact Sheet. http://www.un.org/esa/forests/pdf/session_documents/unff9/Fact_Sheet_ForestsandPeople.pdf

UN (2012). UN urges achieving sustainable energy for all as International Year kicks off. News Centre, 16 January. http://www.un.org/apps/news/story.asp?Cr=energy&NewsID=40951

UN DESA (United Nations Department of Economic and Social Affairs) (2011). World Population Prospects, the 2010 Revision: Frequently Asked Questions (updated 31 Oct. 2011). http://esa.un.org/wpp/Other-Information/faq.htm#q3

UN Security Council (2011). Security Council, in Statement, says 'Contextual Information' on Possible Security Implications of Climate Change Important when Climate Impacts Drive Conflict. News and Media Division, 20 July. http://www.un.org/News/Press/docs/2011/sc10332.doc.htm

UNEP (United Nations Environment Programme) (2010). A Preliminary Assessment. http://www.unep.org/publications/ebooks/emissionsgapreport/pdfs/GAP_REPORT_SUNDAY_SINGLES_LOWRES.pdf

UNEP (2011a). Livelihood Security: Climate Change, Migration and Conflict in the Sahel. Produced through a technical partnership between UNEP, the International Organization for Migration (IOM), the Office for the Coordination of Humanitarian Affairs (OCHA) and the UN University. http://postconflict.unep.ch/publications/UNEP_Sahel_EN.pdf

UNEP (2011b). Decoupling Natural Resource Use and Environmental Impacts from Economic Growth. Report of the Working Group on Decoupling to the International Resource Panel. Fischer-Kowalski, M., Swilling, M., von Weizsäcker, E.U., Ren, Y., Moriguchi, Y., Crane, W., Krausmann, F., Eisenmenger, N., Giljum, S., Hennicke, P., Romero Lankao, P., Siriban Manalang, A. and Sewerin, S. (authors). http://www.unep.org/resourcepanel/decoupling/files/pdf/decoupling_report_english.pdf

UNEP (2011c). Biofuels Vital Graphics: Powering a Green Economy. http://www.grida.no/publications/vg/biofuels/

UNEP (2011d). Bridging the Emissions Gap. http://www.unep.org/pdf/UNEP_bridging_gap.pdf

UNEP (2011e). HFCs: A Critical Link in Protecting the Climate and the Ozone Layer. http://www.unep.org/dewa/Portals/67/pdf/HFC_report.pdf

UNEP (2011f). Towards a Green Economy: Pathways to Sustainable Development and Poverty Eradication. http://www.unep.org/greeneconomy/Portals/88/documents/ger/ger_final_dec_2011/Green%20EconomyReport_Final_Dec2011.pdf

UNEP (2011g). Environmental Assessment of Ogoniland. http://postconflict.unep.ch/publications/OEA/UNEP_OEA.pdf

UNEP (2011h). *UNEP Year Book 2011: Emerging Issues in Our Global Environment*. http://www.unep.org/yearbook/2011h/

UNEP (2012). 21 Issues for the 21st Century: Result of the UNEP Foresight Process on Emerging Environmental Issues. Alcamo, J. and Leonard, S.A. (eds.).

UNEP, IEA and Oeko-Institut (2011). *The Bioenergy and Water Nexus*. http://www.unep.org/pdf/Water_Nexus.pdf

UNEP and IWMI (International Water Management Institute) (2011). *An Ecosystem Services Approach to Water and Food Security*. http://www.iwmi.cgiar.org/topics/ecosystems/PDF/Synthesis_Report-An_Ecosystem_Services_Approach_to_Water_and_Food_Security_2011_UNEP-IWMI.pdf

UNEP and WMO (World Meteorological Organization) (2011). *Integrated Assessment of Black Carbon and Tropospheric Ozone*: Summary for Decision Makers. http://www.unep.org/dewa/Portals/67/pdf/Black_Carbon.pdf

UNHCR (United Nations High Commissioner for Refugees) (2010). *Beyond Emergency Relief: Longer-term trends and priorities for UN agencies in Darfur*. http://reliefweb.int/sites/reliefweb.int/files/resources/6AAEE62B2DC5EE8D852577AE0071F8E7-Full_Report.pdf

UNISDR (United Nations International Strategy for Disaster Risk Reduction) (2011). Halfway through 2011, estimated economic losses already USD265 billion – DRR needed more than ever, says UN. http://www.unisdr.org/archive/20779

UN-REDD (Reducing Emissions from Deforestation and Forest Degradation in Developing Countries) (2011). *REDD+ and a Green Economy: Opportunities for a mutually supportive relationship*. Sukhdev, P., Prabhu, R., Kumar, P., Bassi, A., Patwa-Shah, W., Enters, T., Labbate, G. and Greenwalt, J. (eds.) UN-REDD Policy Brief No. 1. http://www.unredd.net/index.php?option=com_docman&task=doc_download&gid=6345&Itemid=53

US DOI (United States Department of the Interior) (2011). Proposed Outer Continental Shelf Oil and Gas Leasing Program 2012-2017. http://www.boem.gov/uploadedFiles/Proposed_OCS_Oil_Gas_Lease_Program_2012-2017.pdf

US EIA (United States Energy Information Agency) (2011). World Shale Gas Resources: An Initial Assessment of 14 Regions Outside the United States, 5 April. http://www.eia.gov/analysis/studies/worldshalegas/

US EPA (United States Environmental Protection Agency) (2011). *Plan to study the potential impacts of hydraulic fracturing on drinking water resources*. http://water.epa.gov/type/groundwater/uic/class2/hydraulicfracturing/upload/FINAL-STUDY-PLAN-HF_Web_2.pdf

Ververs, M. (2012). The East African Food Crisis: Did Regional Early Warning Systems Function? *The Journal of Nutrition*, 142(1), 131-133

Wittemyer, G., Daballen. D. and Douglas-Hamilton, I. (2011). Poaching policy: Rising ivory prices threaten elephants. *Nature*, 476, 282-283

WMO (World Meteorological Organization) (2011a). Provisional Statement on the Status of the Global Climate. http://www.wmo.int/pages/mediacentre/press_releases/gcs_2011_en.html

WMO (2011b). The Global Framework for Climate Services. http://www.wmo.int/pages/gfcs/gfcs_en.html

WMO (2011c). *WMO Greenhouse Gas Bulletin*, 7. http://www.wmo.int/pages/mediacentre/press_releases/documents/GHGbulletin.pdf

WWF (World Wildlife Fund) (2012). Record Rhino Poaching in South Africa. Press release, 12 January. http://www.worldwildlife.org/who/media/press/2012/WWFPresitem26351.html

The Benefits of Soil Carbon

Managing soils for multiple economic, societal and environmental benefits

In view of the growing world population, within two decades global demand for food is projected to increase by 50 per cent, demand for water by 35-60 per cent, and demand for energy by 45 per cent. The world's soils are consequently under increasing pressure. Soil carbon plays a vital role in regulating climate, water supplies and biodiversity, and therefore in providing the ecosystem services that are essential to human well-being. Managing soils to obtain multiple economic, societal and environmental benefits requires integrated policies and incentives that maintain and enhance soil carbon. Decisive action needs to be taken to limit soil carbon loss due to erosion and emissions of carbon dioxide and other greenhouse gases to the atmosphere.

The top metre of the world's soils stores approximately 2 200 Gt (billion tonnes) of carbon, two-thirds of it in the form of organic matter (Batjes 1996). This is more than three times the amount of carbon held in the atmosphere. However, soils are vulnerable to carbon losses through degradation (**Figure 1**). They also release greenhouse gases to the atmosphere as a result of accelerated decomposition due to land use change or unsustainable land management practices (Lal 2010a, b).

In the face of further land use intensification to meet global demand for food, water and energy (Foresight 2011), managing soils so that carbon stocks are sustained and even enhanced is of crucial importance if we are to meet near-term challenges and conserve this valuable resource for future generations. Since the 19th century, around 60 per cent of the carbon in the world's soils and vegetation has been lost owing to land use (Houghton 1995). In the past 25 years, one-quarter of the global land area has suffered a decline in productivity and in the ability to provide ecosystem services because of soil carbon losses (Bai et al. 2008).

Agriculture on drained peatland in Central Kalimantan, Indonesia, is leading to huge soil carbon losses. *Credit: Hans Joosten*

Authors: Reynaldo Victoria (chair), Steven Banwart, Helaina Black, John Ingram, Hans Joosten, Eleanor Milne and Elke Noellemeyer. Science writer: Yvonne Baskin

Soil erosion associated with conventional agricultural practices can occur at rates up to 100 times greater than the rate at which natural soil formation takes place (Montgomery 2007). Peatland drainage worldwide is causing carbon-rich peat to disappear at a rate 20 times greater than the rate at which the peat accumulated (Joosten 2009).

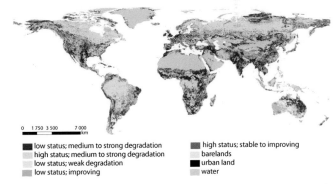

Figure 1: Land degradation can be defined as the reduction in the capacity of the land to provide ecosystem services over a period of time. Pressures from land use can cause degradation. Soil carbon losses are an important form of degradation that can result in loss of productivity and of the ability to provide other ecosystem services. This map shows the status of land in regard to providing capacity for ecosystem services (low, high) and the direction of changes (strong degradation, weak degradation, stable, improving). These global results provide a first indication of pressures and trends at national and regional levels and allow for comparisons to be made between different land uses or geographical regions. *Source: Nachtergaele et al. (2011)*

Box 1: Soil organic matter and soil carbon

Soils are at the heart of the Earth's "critical zone", the thin outer veneer between the top of the tree canopy and the bottom of groundwater aquifers that humans rely on for most of their resources (US NRC 2001, PlanetEarth 2005). They form and continually change over thousands of years, at different rates and along different pathways, as mineral material from the breakdown of rock is colonized by plants and soil biota. This colonization leads to the formation of soil organic matter (SOM) and of soil structure, which controls carbon, nutrient and water cycling (Brantley 2010). Soil carbon exists in both organic and inorganic forms. Soil inorganic carbon is derived from bedrock or formed when CO_2 is trapped in mineral form (e.g. as calcium carbonate). Soil inorganic carbon is far less prone to loss than soil organic carbon (SOC). Although it can dissolve, particularly under acidic conditions, soil inorganic carbon is not susceptible to biodegradation.

SOC is the main constituent of SOM. SOM is formed by the biological, chemical and physical decay of organic materials that enter the soil system from sources above ground (e.g. leaf fall, crop residues, animal wastes and remains) or below ground (e.g. roots, soil biota). The elemental composition of SOM varies, with values in the order of 50 per cent carbon (Broadbent 1953), 40 per cent oxygen and 3 per cent nitrogen, as well as smaller amounts of phosphorus, potassium, calcium, magnesium and other elements as micronutrients. Soil biota (from microbes to earthworms) contribute living biomass to SOM mixing and breaking down the organic matter through physical and biochemical reactions. These biochemical reactions release carbon and nutrients back to the soil, and greenhouse gases such as carbon dioxide (CO_2), nitrous oxide (N_2O) and methane (CH_4) to the atmosphere (**Figure 2**).

Soil management can affect the relative balance of these processes and their environmental impacts. As SOM is broken down, some carbon is mineralized rather rapidly to CO_2 and is lost from the soil. SOM may also be lost through physical erosion. Organic nitrogen contained in biodegrading SOM is transformed to N_2O and other nitrogen oxide (NO_x) compounds. However, some fractions of SOM are not readily degraded. SOC content therefore tends to increase as soil develops undisturbed over time. In water-saturated soils, SOM may even accumulate as thick layers of peat (Beer and Blodau 2007). Organic matter binds to minerals, particularly clay particles, a process that further protects carbon (Von Lützow et al. 2006). Organic matter also provides cohesive strength to soil and improves soil fertility, water movement, and resistance to erosion.

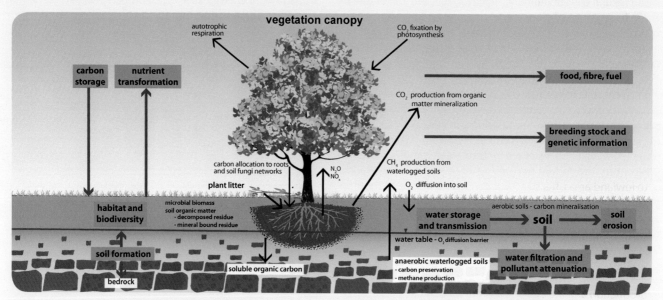

Figure 2: Soil-plant carbon interrelationships and associated ecosystem services. Soils, formed by the action of biota and infiltrating water and solutes on parent rock material, provide ecosystem services as flows of materials (sequestered carbon, water and solutes, plant nutrients, crop biomass) and information encoded in the genetics of soil organisms.

Carbon storage and other vital soil ecosystem services

Scientists have characterized many thousands of different soils. Each has a distinctive composition of minerals, living organisms, organic matter, water and gases (WRB 2006, FAO et al. 2009). Soils are formed over thousands of years as rock is broken down and colonized by plants and soil biota, leading to the formation of soil organic matter (SOM). While SOM is primarily carbon, it also contains nutrients essential for plant growth such as nitrogen, phosphorus, sulphur and micronutrients **(Box 1)**. Organisms in the soil food web decompose SOM and make these nutrients available (Brussaard et al. 2007). The rate of SOM decomposition and turnover mainly depends upon the interplay between soil biota, temperature, moisture and a soil's chemical and physical composition (Taylor et al. 2009).

Soils' use and value are commonly associated with agriculture, but they are also of basic importance to the provision of many other ecosystem services **(Box 2)**. The amount and dynamics of soil carbon are major determinants of the quantity and quality of these services. Ecosystem services are generally divided into four categories:

- Supporting services: These services underpin the delivery of all other services and the benefits that humans obtain from the natural environment. Soil organic matter is a key attribute which influences soils' capacity to support ecosystem services. The inherent characteristics of soils (e.g. soil fertility, soil biodiversity, the capacity to capture, retain and transport water or carbon or to form and release greenhouse gases) are largely determined by the ability of different soils to form and break down soil organic matter.
- Regulating services: Globally, about 75 billion tonnes of soil per year is removed by wind and water erosion (Wachs and Thibault 2009). SOM promotes resistance to erosion of soils and helps regulate flooding by increasing infiltration, reducing runoff and slowing water movement from upland to lowland areas. It also reduces releases of agrochemicals, pathogens and contaminants to the environment by aiding their retention and decomposition (Burauel and Baßmann 2005). Soils have an essential role in climate regulation since soil carbon is the terrestrial biosphere's largest carbon reservoir (Batjes and Sombroek 1997).
- Provisioning services: Soils are the basis of food and fibre production and are of vital importance to recharging water supplies. SOM is necessary to both these services because it influences nutrient and water availability and soil structure. It also increases resilience to climate change by helping protect plants and the environment against water stress and excess water. Carbon-rich peat soils have been a source of fuel throughout history. Today they provide growing media for gardeners, horticulturists and industry.
- Cultural services: From ancient times, human cultures have been strongly affected by the ways they use and manage soils. The character and carbon content of these soils have influenced the nature of landscapes and the environments in which diverse cultures have developed and thrived. SOM also helps soils to retain traces of past cultures and climates and to preserve archaeological remains.

Water infiltration is reduced in degraded soils. Therefore, in such soils less rainfall infiltrates to recharge soil and groundwater and more is lost to evaporation and runoff. *Credit: Elke Noellemeyer*

Water-saturated peatlands can conserve archaeological remains virtually forever. The mummified body of the Tollund Man, who lived 2 500 years ago, was found in 1950 in a peat bog in Denmark. *Credit: Cochyn*

> **Box 2:** Soils are of basic importance to the delivery of many interrelated ecosystem services. *Source: MEA (2005), Black et al. (2008)*
>
> **Supporting services:** nutrient cycling, water release/retention, soil formation, habitat for biodiversity, exchange of gases with the atmosphere, degradation of complex materials
>
> **Regulating services:** carbon sequestration, greenhouse gas emissions, water purification, natural attenuation of pollutants
>
> **Provisioning services:** food and fibre production, water availability, platform for construction
>
> **Cultural services:** protection of archaeological remains, outdoor recreational pursuits, landscapes, supporting habitats
>
>
>
> Credit: Elke Noellemeyer Credit: Márton Bálint Credit: Anja Leide Credit: Kevin Bacher, NPS

Managing soil carbon for multiple benefits is key to its sustainable use. Trade-offs among the benefits that ecosystem services provide arise when soil management is focused on a single ecosystem service. For instance, using drained peatlands for biomass production greatly diminishes soil carbon stocks, degrades native habitats and alters the peatlands' capacity to provide climate-regulating services. In contrast, soil carbon can be managed to enhance a range of ecosystem services. Increasing the SOM of degraded soils can simultaneously boost agricultural productivity, sequester CO_2 whose emissions might otherwise exacerbate climate change, and enhance water capture.

What determines the global distribution of soil carbon?

The worldwide distribution of SOC reflects rainfall distribution, with greater accumulations of carbon in more humid areas (**Figure 3**). Most SOC is found in the northern hemisphere, which contains more land mass in humid climates than the southern hemisphere. Temperature plays a secondary role in global SOC distribution. This is illustrated by the occurrence of deep peat deposits in both tropical and polar humid areas.

Within climatic zones the amount of SOC is determined by soil moisture, which in turn is influenced by relief, soil texture and clay type. High soil water content tends to conserve SOM because reduced oxygen availability in wet soils slows the decomposition of SOM by soil microbes. Drier and well-aerated soils promote more rapid decomposition and accumulate less SOM. Where soil oxygen, soil moisture levels and nutrient status are sufficient, higher temperatures accelerate biological processes such as biomass production and decomposition, and therefore SOC dynamics (Batjes 2011). That is why draining peatlands provokes a rapid oxidation of stored SOM and releases large amounts of CO_2 to the atmosphere, especially in warmer climates. Similarly, conversion of natural grasslands or forests to tilled soils breaks up soil aggregates, produces better aeration, and thus increases the decomposition of SOM and releases of CO_2, with higher rates occurring in warm climates. Scientists have shown that in arable agriculture "no-till" land management reduces carbon losses and enhances the potential for carbon sequestration (**Box 3**).

The carbon content of soils under different land cover types varies substantially (**Figure 4**). The soils of savannahs are relatively low in SOC, but the carbon stocks of savannah soils are significant globally due to the large land area covered by this biome. In contrast, peatlands cover only 3 per cent of the global land area but contain almost one-third of global soil carbon, making them the most space-effective carbon store among all terrestrial ecosystems. Drained and degrading peatlands, which occupy 50 million ha worldwide (0.3 per cent of the global land area), produce more than 2 Gt of CO_2 emissions annually – equivalent to 6 per cent of all global anthropogenic CO_2 emissions (Joosten 2009).

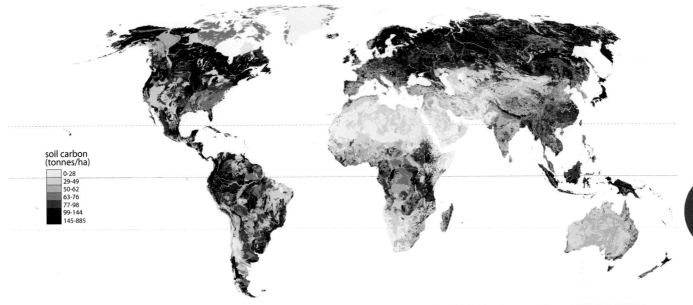

Figure 3: Organic soil carbon to a depth of 1 metre in tonnes per ha. Data are derived from the Harmonized World Soil Database v1.1. *Source: UNEP-WCMC (2009)*

Modelling, measuring and monitoring

Ways to estimate soil carbon stocks and fluxes at scales ranging from field to global continue to be developed (Bernoux et al. 2010, Hillier et al. 2011). The lack of adequate methodologies and approaches has been one of the main barriers to accounting for the significant mitigation effects that land management projects can have. This is important in the case of projects whose purpose is to sequester carbon in biomass or soils, or those that include sequestration as a co-benefit of reducing rural poverty or addressing food security. The Global Soil Mapping initiative will help provide a globally consistent set of soils data that are geographically continuous, scalable and include uncertainty estimates (Global Soil Mapping 2011). These data need to be

---

Box 3: Effects of no-tillage land management in Argentina and Brazil

In Argentina, which is currently experiencing agricultural expansion, no-tillage ("no-till") land management has proven a viable alternative to conventional cultivation that involves working the soil with ploughs and harrows several times before seeding. Along with enhanced benefits from better water retention and infiltration and erosion prevention, small but significant increases in SOC stocks have been achieved where farmers changed to no-till systems (Alvarez and Steinbach 2009, Fernández et al. 2010).

In Brazil, changes in crop production practices have also had significant effects on soil carbon stocks. Conversion to no-till in soybean, maize and related crop rotation systems has resulted in a mean SOC sequestration rate of 0.41 tonnes per hectare per year. Pastures also have potential for soil carbon sequestration when integrated with arable agriculture (rotations), with

Soybean fields in the semi-arid Argentinian pampa. After 15 years no-till (right) carbon levels at a 0-20 cm soil depth were 15.8 tonnes per ha compared to 13.8 tonnes per ha under conventional cultivation (left).
Source: Fernández et al. (2010), Credit: Elke Noellemeyer

the added benefit of increasing agricultural production (De Figueiredo and La Scala 2011, La Scala et al. 2011).

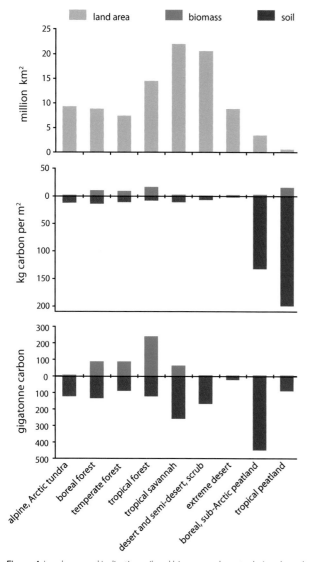

Figure 4: Land area and indicative soil and biomass carbon stocks in selected biomes, land use and land cover types. Soil carbon stocks refers to the upper metre of soil except in the case of peatlands. The three forest biomes and tundra include some peatland carbon in their overall soil carbon stocks. *Source: Adapted from Gorham (1991), ORNL (1998), Verwer and Van de Meer (2010) and Page et al. (2011)*

enhanced by contemporary field measurements (such as those supported in Africa by The Bill & Melinda Gates Foundation and the Africa Soil Information Service) and by progressive soil monitoring (Africa Soil Information Service 2011, Gates Foundation 2011).

There is also a critical need to develop universally agreed and reproducible field and laboratory methods for measuring, reporting and verifying (MRV) changes in soil carbon over time.

Box 4: The Carbon Benefits Project

Land use, land-use change and forestry (LULUCF) activities can provide a relatively cost-effective way to offset emissions through increasing removals of greenhouse gases from the atmosphere (e.g. by planting trees or managing forests) or through reducing emissions (e.g. by curbing deforestation) (UNFCCC 2012). However, it is often difficult to estimate greenhouse gas removals and emissions resulting from LULUCF activities. The UNEP-Global Environment Facility's Carbon Benefits Project: Modelling, Measurement and Monitoring has developed a set of scientifically rigorous, cost-effective tools to establish the carbon benefits of sustainable land management interventions. These tools are designed to estimate and model carbon stocks and flows and greenhouse gas emissions under present and alternative management, and to measure and monitor carbon changes under specified land use (**Figure 6**).

An online toolset can be applied to projects involving soil services and natural resources management (e.g. forestry, agroforestry, agriculture and pasture management) in all climate zones. The modelling system enables projects to assess sources and sinks of CO_2 and other greenhouse gases at all points in a project cycle. The measurement system uses a combination of remotely sensed observations, ground calibration and web-enabled geographic information systems. It also provides estimates of CH_4 and N_2O dynamics based on direct field flux measurements.

Such approaches could allow for large area landscape assessments of above- and below-ground carbon for policy mechanisms whose purpose is to mitigate climate change through Reducing Emissions from Deforestation and forest Degradation (REDD) in developing countries. Greenhouse gas removals and emissions through afforestation and reforestation (A/R) since 1990 could be accounted for in meeting the Kyoto Protocol's emission targets under certain rules (UNFCCC 2012). Parties could also select additional human-induced LULUCF activities, such as grazing land management, cropland management, forest management and revegetation, for which such tools could be useful.

Carbon measurement is currently being debated at several levels to correctly address carbon markets, for instance in the agricultural and forest sectors. Improved estimations of SOM, carbon stocks and fluxes could greatly help scientists to monitor and predict ecosystems' response to climate change, as well as aiding policy makers when they take land use and management decisions and assisting land managers to gain better access to carbon markets

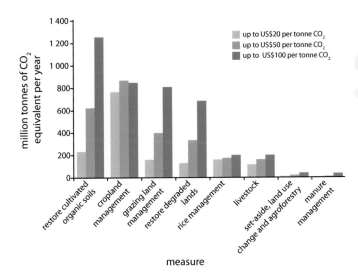

Figure 5: The effect of market prices on the efficacy of management measures to increase soil carbon. *Source: Adapted from Smith et al. (2007)*

(Smith et al. 2007, Ravindranath and Ostwald 2008, Milne et al. 2010, FAO 2011, Schmidt et al. 2011). When appropriate measurement and incentive mechanisms are in place, market prices can have a significant effect on the efficacy of management measures to increase soil carbon **(Figure 5)**. The Carbon Benefits Project is developing a tool that aims to address the issue of estimating carbon benefits, along with measuring and monitoring the effect of land management interventions **(Box 4)**.

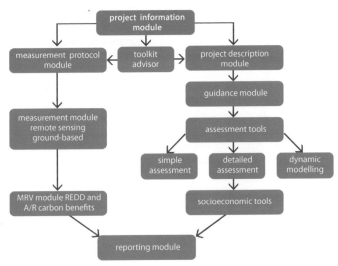

Figure 6: The concept behind the Carbon Benefits Project online toolset. *Source: CBP (2012)*

The vulnerability of soil carbon stocks to human activities

Soil carbon stocks are highly vulnerable to human activities. They decrease significantly (and often rapidly) in response to changes in land cover and land use such as deforestation, urban development and increased tillage, and as a result of unsustainable agricultural and forestry practices. SOC may also be increased (although much more slowly) by afforestation and other activities that decrease the breakdown of SOM (e.g. minimum tillage, perennial pastures, designation of protected areas). Practices that add more organic matter to the soil, such as composting or adding manure, may only improve the carbon balance of one site while diminishing that of another. Climate change is expected to have significant impacts on soil carbon dynamics (Schils et al. 2008, Conant et al. 2011). Rising atmospheric CO_2 levels could increase biomass production and inputs of organic materials into soils. However, increasing temperatures could reduce SOC by accelerating the microbial decomposition and oxidation of SOM, especially in thawing permafrost soils. Experts are concerned that if permafrost thaws, enormous amounts of carbon might be released into the air, greatly intensifying global warming (Schuur and Abbott 2011). Although the magnitude of this effect remains highly uncertain, recent estimates of frozen soil carbon are huge. According to some scientists, some 18.8 million km^2 of northern soils hold about 1 700 billion tonnes of organic carbon (Tarnocai et al. 2009).

Current scientific knowledge of how local soil properties and climatic conditions affect soil carbon stock changes and carbon fluxes is insufficient and conflicting (Tuomi et al. 2008, Conant et

Permafrost thaws could result in releases of enormous amounts of carbon to the air. *Credit: Hans Joosten*

al. 2011, Falloon et al. 2011). Further study will be needed to enable more accurate predictions of the impacts of climate change on soils, soil carbon and associated ecosystem services at scales relevant to local management, as well as to national carbon inventories.

The current rate of change in SOC is mainly attributable to worldwide land use intensification and the conversion of new land for food and fibre production. Modern industrialized crop production relies on monocultures of highly efficient cash crops, which generally create a negative carbon budget. Alternative uses of crop residues for fodder, fuel or industrial applications exacerbate this trend of decreasing carbon return to the soil. Crop type also plays a part. Soybean monocultures, which have recently spread widely, accelerate SOC losses because their scant crop residues provide less cover to protect soils from wind and water erosion, are highly labile and are rapidly oxidized to CO_2. Intensive animal production systems that harvest all plant biomass also reduce SOC stocks compared to traditional grazing systems, which only partially remove plant biomass. The overall impact of such intensification is that, although rates of carbon emissions from much of the world´s arable land remain low, large areas are experiencing decreases in SOC stocks. Globally, therefore, soils under intensive agricultural use can be considered an important source of atmospheric CO_2 and other greenhouse gases (Janzen 2006, Powlson et al. 2011).

Intensive land uses are also expanding into areas where SOC stocks are less resilient or soil conditions are marginal for agriculture. For example, semi-arid savannahs and grasslands, tropical rainforests and peatlands are all being converted to arable land at an increasing rate. While temperate humid grasslands lose about 30 per cent of their SOC after 60 years of cultivation (Tiessen and Stewart 1983), soil carbon stocks in semi-arid environments can decrease by 30 per cent in less than five years when native vegetation or pastures are converted to cropland (Zach et al. 2006, Noellemeyer et al. 2008). Pastures established on cleared Amazon rainforest emit between 8 and 12 tonnes of carbon per hectare (Fearnside and Barbosa 1998, Cerri et al. 2007). Cultivation of tropical forest soils causes losses

---

Box 5: Paludiculture: sustainable cultivation of peatlands

Harvesting biomass in Poland's Biebrza peatland. *Credit: Lars Lachmann, BMBF-VIP-Project*

Current use of drained peatlands on only 0.3 per cent of the Earth's surface is responsible for a disproportionate 6 per cent of global human-generated CO_2 emissions (Joosten 2009). Drained peatlands are increasingly used to produce biomass fuels such as palm oil in southeast Asia, sugarcane in Florida, maize and miscanthus in temperate Europe, and wood in parts of Scandinavia. This type of cultivation causes much greater CO_2 emissions than those saved by replacing fossil fuels with these biomass fuels (Couwenberg 2007, Sarkkola 2008, Wicke et al. 2008, Couwenberg et al. 2010).

Paludiculture (*palus* is the Latin word for swamp) is biomass cultivation on wet and rewetted peatlands (Wichtmann et al. 2010). It offers an innovative alternative to conventional peatland agriculture and silviculture. Paludicultures can contribute to climate change mitigation in two ways: by reducing greenhouse gas emissions through rewetting of drained peatland soils; and by replacing fossil resources with renewable biomass alternatives.

Rewetting peatlands is generally beneficial for biodiversity since strongly degraded peatlands are ecological deserts. Regular harvesting of cultivated biomass on undrained or rewetted peatlands keeps vegetation short, reduces nutrient levels, and allows less competitive native species to (re-)establish. An example is the aquatic warbler, a fen flagship species that survives only in harvested wetlands (Tanneberger et al. 2009).

Paludicultures offer a sustainable future for managed peatlands as productive land. Although special wetland-adapted harvest machinery is required, winter-harvested reed from paludicultures in northeast Germany can fully compete with miscanthus or straw grown on mineral soils.

Agriculture on drained peatland, such as here in Central Kalimantan, Indonesia, leads to very large carbon losses. *Credit: Hans Joosten*

of more than 60 per cent of original SOC stocks in just a few years (Brown and Lugo 1990).

Tropical peatlands converted to cropland or plantations are another hotspot for carbon emissions **(Box 5)**. Draining peat soils to introduce commercial production systems in tropical environments causes ongoing losses of up to 25 tonnes of carbon per hectare per year (Jauhiainen et al. 2011), while in boreal peatlands emissions from cropland are around 7 tonnes per hectare per year (Couwenberg 2011).

Consequences of soil carbon loss and the potential for soil carbon gain

Soil carbon losses not only result in higher atmospheric CO_2 concentrations through accelerated soil carbon oxidation, but also in a general loss of soil functioning and soil biodiversity. Less SOM, leads to decreased cohesion between soil particles, which increases the susceptibility of soil to water or wind erosion, accelerates losses of bulk soil, and alters nutrient and water cycling. Degradation of soil structure reduces the soil volume for water storage and soil permeability for drainage. In turn, this can lead to greater volumes of overland flow, which exacerbates flooding and reduces groundwater recharge during rain events. Reduced groundwater recharge aggravates water shortages and drought conditions. Another consequence of soil carbon loss is the loss of soil nutrients. These include nutrient elements within the SOM, as well as inorganic nutrients such as phosphorus and potassium that bind to mineral surfaces. Because of SOM's role in forming aggregates, loss of SOM can reduce soil cohesion and allow the breakup of these aggregates (Malamoud et al. 2009). This increases the potential to lose bound clays and other minerals, either through bulk erosion or through colloid transport as water percolates through the soil profile.

In view of the many benefits of soil carbon, priority should be given to maintaining SOC levels in soils and, wherever possible, increasing these levels. Soil carbon gains can be achieved in two ways: first, by applying management strategies (including "set-aside" of land where this is socially and economically feasible) and technologies that reduce losses of existing soil carbon (this is particularly important in the case of dryland soils and natural grasslands or savannahs); and second, by applying sustainable management techniques that increase the levels of carbon in soils, particularly degraded agricultural soils **(Box 6)**.

SOC losses can be reduced by minimizing oxidation of SOC in the soil profile and by reducing soil removal (e.g. removal of peat for fuel or horticultural use, or of soil for construction). In the case of mineral soils, which are typical of major cropping regions, reducing tillage can minimize soil carbon losses. In addition, carbon in the soil surface can be protected through practices that control erosion, such as shelter belts, contour cultivation and cover crops. In peat soils the naturally high carbon density can be preserved by maintaining wet saturated conditions, rather than by draining the peatlands to accommodate forestry or cropping such as oil palm plantations. In already degraded peat soils, raising water levels through drain-blocking can reduce further oxidation and contribute to maintaining and restoring carbon levels (Tanneberger and Wichtmann 2011). However, care should be taken not to inundate labile organic matter, such as fresh crop residues, since waterlogged conditions can lead to anaerobic decomposition that can produce large amounts of methane (CH_4), a potent greenhouse gas (Couwenberg et al. 2011).

Increasing SOC levels, on the other hand, can be achieved by increasing carbon inputs to soils. In the case of managed soils, this can be done by increasing the input and retention of above-ground biomass. Plants also allocate a significant portion of carbon below ground via their roots. This supports the soil biota in the rooting zone and, in turn, facilitates plant nutrient uptake, resulting in improved crop productivity and further increasing flows of carbon into the soil. Thus, sustainable land management for enhanced SOC levels is based on: optimal plant productivity (crop selection, appropriate soil nutrient management, irrigation); minimal losses of organic matter in soil (reduced tillage, erosion control, cover crops); and high carbon returns to the soil (i.e. leaving post-harvest crop residues or importing organic matter such as animal manures, biochar and domestic or industrial wastes, after consideration of the potential risks associated with using these materials).

Box 6: Strategies for maintaining and increasing carbon stocks in three major land use systems

Carbon stocks can be enhanced by ensuring that carbon inputs *to* the soil are greater than carbon losses *from* the soil. Different strategies are required to achieve this objective, depending on land use, soil properties, climate and land area.

Grasslands

The improvement of soil carbon in grasslands offers a global greenhouse gas mitigation potential of 810 Mt of CO_2 (in the period up to 2030), almost all of which would be sequestered in the soil (Conant et al. 2001, Ravindranath and Ostwald 2008). Overstocking of grazing animals can lead to the degradation of grasslands, increased soil erosion, depletion of SOC and increased soil greenhouse gas emissions. It should therefore be avoided. Activities that improve soil carbon in grasslands may include the following:

- Adding manures and fertilizers can have a direct impact on SOC levels through the added organic material and indirect impacts though increasing plant productivity and stimulating soil biodiversity (e.g. with earthworms that help degrade and mix the organic material). Fertilizer use can, however, result in N_2O emissions.
- Revegetation, especially using improved pasture species and legumes, can increase productivity, resulting in more plant litter and underground biomass, which can augment the SOC stock.
- Irrigation and water management can improve plant productivity and the production of SOM. These gains, however, should be set against any greenhouse gas emissions associated with energy used for irrigation, nutrient leaching affecting water quality, and the risks of evaporative salt deposits adversely impacting soil fertility.

Croplands

Techniques for increasing SOC in the agricultural sector include the following (Altieri 1995):

- Mulching can add organic matter. If crop residues are used, mulching also prevents carbon losses from the system. However, in flooded soils mulching can increase CH_4 emissions.
- Reduced or no tillage avoids the accelerated decomposition of organic matter and depletion of soil carbon that occur with intensive tillage (ploughing). Reduced tillage also prevents the break-up of soil aggregates that protect carbon.
- Judicious use of animal manure or chemical fertilizers can increase plant productivity and thus SOC, although adding excess nutrients can also increase losses of SOC as greenhouse gas emissions. All fertilizer additions must also consider the greenhouse gas costs of production and transport set against higher crop yields, which may offset demands for production on marginal land and farm-to-market transport.
- Rotations of cash crops with perennial pastures, and (in certain climates and farming systems) the use of cover crops and green manures have the potential to increase biomass returned to the soil and can therefore increase soil carbon stocks.
- Using improved crop varieties can increase productivity above and below ground, as well as increasing crop residues, thereby enhancing SOC.
- Site-specific agricultural management can reduce the risk of crop failure and thus improve an area's overall productivity, improving carbon stocks.
- Integration of several crops in a field at the same time can increase organic material, soil biodiversity and soil health, as well as increasing food production, particularly for subsistence farmers.

Forested lands and tree crops

Forests have considerable potential to reduce greenhouse gas emissions to the atmosphere by storing large stocks of carbon both above and below ground. Strategies for realizing this potential include:

- Protection of existing forests will preserve current soil carbon stocks.
- Reforesting degraded lands and increasing tree density in degraded forests increase biomass density, and therefore carbon density, above and below ground.
- Trees in croplands (agroforestry) and orchards can store carbon above and below ground and even reduce fossil fuel emissions if they are grown as a renewable source of firewood.

Millet field and millet storage bins in Niger. *Credit: Curt Reynolds*

Restoration of hydrology and plant communities in agricultural fields may result in net soil carbon sequestration, North Carolina, USA. *Credit: J.L. Heitman*

In all cases, the success of strategies to increase soil carbon stocks will depend on the intrinsic capacity of a particular soil (e.g. mineral composition, clay content) and local soil formation conditions (e.g. climate, slope) as well as the nature of land use and management. To ensure that net SOC changes are real, greater emphasis needs to be placed on the assessment of SOC to depth alongside greenhouse gas emissions. Research is also needed to characterize the intrinsic SOC holding capacity of different soils in order to better target investment in management practices – that is, to compare current baseline SOC stocks and fluxes against potential ones under alternative management.

The way forward: managing soil carbon for multiple benefits

The world is experiencing rapid and unprecedented changes in land use, driven by increasing demand for food, water, energy and space for living (Verburg et al. 2011). Historically, the demand for food and fibre has been met by converting natural and semi-natural habitats to cropland in order to cultivate fertile soils with significant soil carbon stocks. As this demand grows in the future, cropping intensity will need to increase as less land becomes available for conversion to agriculture (Bruinsma 2003). Such land conversions have major implications for soil carbon stocks (Smith et al. 2010).

If current trends continue, there will be rapid losses of soil carbon to the atmosphere in years to come – not only exacerbating climate change, but also increasing the extent of global soil degradation and diminishing a wide range of vital ecosystem services. The consequences of further losses of soil carbon may take several decades to become obvious, by which time they could be difficult or expensive to address.

Soil carbon stocks vanish rapidly as a result of land use change and unsustainable management, while replenishing them is slow and requires significant investment. Positive actions can be taken now to avoid SOC losses by protecting soil carbon stocks and promoting sustainable practices that enhance SOC. Comprehensive accounting of social, economic and environmental costs and benefits can help ensure wide understanding of the local to global implications of land use and management changes that will maintain, enhance or degrade SOC.

Opportunities exist at the global, regional and local levels to enhance soil carbon and avoid losses of this resource. The challenge is to develop and implement planning processes, policies and incentive mechanisms that balance pressures on the soil from contrasting and (at times) conflicting demands for food, fibre and fuel crops, climate regulation, water, biodiversity conservation, living space and other benefits. In some locations, mechanisms will be needed to protect soils that are important soil carbon stores, such as peatlands and tundra, as an alternative to other uses such as agricultural or forestry expansion. However, in many cases multiple economic, societal and environmental benefits can be obtained on the same land through effective management of soil carbon.

There are examples worldwide of how multiple benefits can be derived through effective soil carbon management (UNEP-WCMC 2008, Marks et al. 2009, Kapos et al. 2010, Reed et al. 2010, Watson 2010). For instance, The World Bank's BioCarbon Fund is providing the Kenya Agricultural Carbon Project with US$350 000 to pay smallholder farmers to improve their agricultural practices, in order to increase both food security and soil carbon sequestration (World Bank 2010). In parallel, the Great Green Wall initiative is a massive afforestation project to create a 15 km wide strip of trees and other vegetation along a 7 000 km transect of the African continent from Senegal to Djibouti (Bellefontaine et al. 2011). The objectives of this project include carbon sequestration, stabilization of soils, conservation of soil moisture and support for agriculture. Similar approaches are being monitored in China to assess whether they can sustainably reverse land degradation in arid regions (Bai and Dent 2009).

Proven technologies and management options are available for SOC conservation and enhancement, but whether they can be widely applied will be determined by the policies and incentives that encourage their use. Currently, the value of soil carbon (and soils in general) is rarely considered across sectors. The perceived benefits of soil carbon often reflect only the primary demands of a particular land use such as food production. In some parts of the world, land management strategies and policies are already

Soil policy integration can deliver multiple economic, societal and environmental benefits. *Credit: Clean Seed Capital*

in place to balance production pressures against other societal goals such as enhancing biodiversity or improving water quality. However, none were designed specifically to optimize soil management for multiple carbon benefits and associated ecosystem services. For instance, organic inputs to agricultural soils are generally targeted at increasing soil fertility, although this practice can also reduce soil erosion, achieve soil carbon sequestration and add resilience to farming systems.

There is a clear opportunity to use existing mechanisms, individually and in combination, to encourage active management of soil carbon and therefore expand the range of potential benefits. Where strategies to apply such mechanisms do not exist, there is an opportunity to design new strategies that take into account the multiple benefits of soil carbon management. Various global efforts and policy options exist that could be augmented to achieve wider benefits from SOC gains (**Box 7**).

Ultimately, these global agreements and policies could be linked in ways that encourage delivery of multiple benefits from soil carbon. In September 2011, the Food and Agriculture Organization of the United Nations (FAO) began planning with other UN agencies, including UNEP, for the establishment of a Global Soil Partnership to support and facilitate joint efforts towards sustainable management of soil resources for food security and for climate change adaptation and mitigation (FAO 2012). National and local regulations and incentives can also be used to promote improved soil carbon management for multiple benefits, with respect to existing land uses as well as restoration of degraded soils.

Box 7: Global policy options to achieve soil carbon benefits

- International climate change efforts to reduce the intensity of global warming (e.g. the United Nations Framework Convention on Climate Change) could indirectly reduce soil carbon losses by halting the acceleration of SOC losses in highly organic soils of the tundra and elsewhere, precluding the expansion of intensified land use in areas where current climatic conditions limit cultivation, such as mountainous regions, and promoting SOC gains in agricultural soils.
- Actions to address land degradation (e.g. under the United Nations Convention to Combat Desertification) could reduce carbon losses by encouraging soil conservation measures to prevent soil erosion and increase carbon stocks in affected areas, as well as by promoting practices that enhance soil organic matter (SOM) to restore degraded soils.
- Trade policies (e.g. through the World Trade Organization) could promote the market benefits from increasing soil carbon (e.g. through better prices for products derived from sustainable, carbon-friendly production systems identified by labelling) or counter losses of soil carbon from the expansion of particular land uses or crop types into vulnerable areas (e.g. through controls on the marketing of products that come from drained peatlands or the conversion of tropical rainforest).
- Global agreements could include tradable carbon or other credits (e.g. "green water") for soils as a mechanism to manage soil resources in order to obtain environmental, social and economic benefits both on- and off-site (Tanneberger and Wichtmann 2011). Widespread adoption of soil organic carbon (SOC) management strategies will be influenced by the stability and level of the market price for SOC, as well as by access to financial mechanisms and incentives and by local issues such as land tenure. Carbon credits will only be effective if SOC sequestration can be adequately monitored and evaluated, and if long-term social and environmental impacts are adequately considered alongside short-term economic benefits.
- Conservation policies whose purpose is to halt biodiversity loss and protect ecosystems can also protect soil carbon stocks (e.g. when peatlands are rewetted or above-ground vegetation is restored) (Bain et al. 2011). The Convention on Biological Diversity and the Ramsar Convention on Wetlands focus on the protection and conservation of designated areas. An international mechanism to protect the world's soil heritage exists under the World Heritage Convention. Its implementation would serve to improve the protection and management of soil resources, including soil carbon.

Such mechanisms could include:
- Land use planning that excludes vulnerable soils from land uses that lead to SOC losses.
- Promotion of management to protect and enhance SOM as an essential element of good soil and environmental quality.
- Regulations and guidelines on limiting emissions of greenhouse gases to the atmosphere, releases of soil carbon, nitrate and other contaminants to surface and groundwater, and drainage of carbon-rich soils.
- Promotion of sources of plant nutrients (e.g. cover crops, legumes, plant growth-promoting "bio-effectors") that enhance SOC stocks.
- Financial incentives such as payments for carbon storage, flood control, improvement of water quality, conservation of soil biodiversity and other ecosystem services.
- Technical advisory systems (extension services) for agriculture and forestry that address the full range of ecosystem services that are supported by soils.

Soil carbon is easily lost but difficult to rebuild. Because it is central to agricultural productivity, climate stabilization and other vital ecosystem services, creating policy incentives around the sustainable management of soil carbon could deliver numerous short- and long-term benefits. Such policy incentives would need to target better allocation of soil resources to different land uses and management practices than has been the case under current policies targeted at supplying individual ecosystem services. Carefully crafted integrated policies could also avoid creating financial incentives that establish new conflicts or trade-offs involving soil carbon.

A new focus at all levels of governance on managing soils for multiple benefits by managing soil carbon effectively would constitute a significant step towards meeting the need for ecosystem services to support the world population in 2030 and beyond.

The aquatic warbler (*Acrocephalus paludicola*) is globally threatened and survives only in harvested, carbon rich wetlands. *Credit: Franziska Tanneberger*

References

Africa Soil Information Service (2011). Africa Soil Information Service Labs: Data Analysis. http://africasoils.net/labs/data-analysis-2/

Altieri, M.A. (1995). *Agroecology: The Science of Sustainable Agriculture*. Westview Press, Boulder, Colorado, USA

Álvarez, R. and Steinbach, H. (2009). A review of the effects of tillage systems on some soil physical properties, water content, nitrate availability and crop yields in the Argentine Pampas. *Soil & Tillage Research*, 104, 1-15

Bai, Z.G. and Dent, D.L. (2009). Recent land degradation and improvement in China. *Ambio*, 38, 150-156

Bai, Z.G., Dent, D.L., Olsson, L. and Schaepman, M.E. (2008). Proxy global assessment of land degradation. *Soil Use and Management*, 24, 223-234

Bain, C.G., Bonn, A., Stoneman, R., Chapman, S., Coupar, A., Evans, M., Gearey, B., Howat, M., Joosten, H., Keenleyside, C., Labadz, J., Lindsay, R., Littlewood, N., Lunt, P., Miller, C.J., Moxey, A., Orr, H., Reed, M., Smith, P., Swales, V., Thompson, D.B.A., Thompson, P.S., Van de Noort, R., Wilson, J.D. and Worrall, F. (2011). IUCN UK Commission of Inquiry on Peatlands. IUCN UK Peatland Programme, Edinburgh, UK

Batjes, N.H. (1996). Total carbon and nitrogen in the soils of the world. *European Journal of Soil Science*, 47, 151-163

Batjes, N.H. (2011). Soil organic carbon stocks under native vegetation - revised estimates for use with the simple assessment option of the Carbon Benefits Project system. *Agriculture, Ecosystems & Environment*, 142, 365-373

Batjes, N.H. and Sombroek, W.G. (1997). Possibilities for carbon sequestration in tropical and sub-tropical soils. *Global Change Biology*, 3, 161-173

Beer, J. and Blodau, C. (2007). Transport and thermodynamics constrain belowground carbon turnover in a northern peatland. *Geochimica et Cosmochimica Acta*, 71, 2989-3002

Bellefontaine, R., Bernoux, M., Bonnet, B., Cornet, A., Cudennec, C., D'Aquino, P., Droy, I., Jauffret, S., Leroy, M., Malagnoux, M. and Réquier-Desjardins, M. (2011). The African Great Green Wall project. What advice can scientists provide? *CSFD Topic Briefs – February 2011*. http://www.csf-desertification.org/pdf_csfd/GMV/fiche-A4-GMV-eng.pdf

Bernoux, M., Branca, G., Carro, A., Lipper, L., Smith, G. and Bockel, L. (2010). Ex-ante greenhouse gas balance of agriculture and forestry development programs. *Scientia Agricola*, 67(1), 31-40

Black, H.I.J., Glenk, K, Towers, W., Moran, D. and Hussain, S. (2008). Valuing our soil resource for sustainable ecosystem services. Scottish Government Environment, Land Use and Rural Stewardship Research Programme 2006-2010. http://www.programme3.net/soil/p3-soilsposter-2.pdf

Brantley, S. (2010). Weathering Rock to regolith. *Nature Geoscience*, 3, 305

Broadbent, F.E. (1953). The soil organic fraction. *Advances in Agronomy*, 5, 153-183

Brown, S. and Lugo, A.E. (1990). Effects of forest clearing and succession on the carbon and nitrogen contents of soil in Puerto Rico and *U.S. Plant and Soil*, 124, 53-64

Bruinsma, Jed. (2003). *World Agriculture: Towards 2015/2030, an FAO Perspective*. Earthscan Publications, London

Brussaard, L., de Ruiter, P.C. and Brown, G.G. (2007). Soil biodiversity for agricultural sustainability. *Agriculture, Ecosystems & Environment*, 121, 233-244

Burauel, P. and Baßmann, F. (2005). Soils as filter and buffer for pesticides – experimental concepts to understand soil functions. *Environmental Pollution*, 133, 11-16

CBP (in preparation). Carbon Benefits Project: www.unep.org/ClimateChange/carbon-benefits/cbp_pim

Cerri, C.E.P., Easter, M., Paustian, K., Killian, K., Coleman, K., Bernoux, M., Falloon, P., Powlson, D.S., Batjes, N.H., Milne, E and Cerri, C.C. (2007). Simulating SOC changes in 11 land use change chronosequences from the Brazilian Amazon with RothC and Century models. *Agriculture Ecosystems and Environment*, 122(1), 46-57

Conant, R.T., Paustian, K. and Elliott, E.T. (2001). Grassland management and conversion into grassland: Effects on soil carbon. *Ecological Applications*, 11(2), 343–355

Conant, R.T., Ryan, M.G., Ågren, G.I., Birge, H.E., Davidson, E.A., Eliasson, P.E., Evans, S.E., Frey, S.D., Giardina, C.P., Hopkins, F.M., Hyvönen, R., Kirschbaum, M.U.F., Lavallee, J.M., Leifeld, J., Parton, W.J., Megan Steinweg, J., Wallenstein, M.D., Wetterstedt, J.Å.M. and Bradford, M.A. (2011). Temperature and soil organic matter decomposition rates synthesis of current knowledge and a way forward. *Global Change Biology*, 17, 3392-3404

Couwenberg, J. (2007). Biomass energy crops on peatlands: On emissions and perversions. *IMCG Newsletter*, 3, 12-14

Couwenberg, J. (2011). Greenhouse gas emissions from managed peat soils: Is the IPCC reporting guidance realistic? *Mires and Peat*, 8, 1-10. http://www.mires-and-peat.net/

Couwenberg, J., Dommain, R. and Joosten, H. (2010). Greenhouse gas fluxes from tropical peatlands in south-east Asia. *Global Change Biology*, 16, 1715-1732

Couwenberg, J., Thiele, A., Tanneberger, F., Augustin, J., Bärisch, S., Dubovik, D., Lianshchynskaya, N., Michaelis, D., Minke, M., Skuratovitch, A. and Joosten, H. (2011). Assessing greenhouse gas emissions from peatlands using vegetation as a proxy. *Hydrobiologia*, 674, 67-89

De Figueiredo, E.B. and La Scala Jr., N. (2011). Greenhouse gas balance due to the conversion of sugarcane areas from burned to green harvest in Brazil. *Agriculture, Ecosystems and Environment*, 141, 77-85

Dexter, A.R. (2004). Soil physical quality Part I. Theory, effects of soil texture, density, and organic matter, and effects on root growth. *Geoderma*, 120, 201-214

Falloon P., Jones, C.D., Ades, M. and Paul, K. (2011). Direct soil moisture controls of future global soil carbon changes: An important source of uncertainty. *Global Biogeochemical Cycles*, 25, GB3010

FAO (Food and Agriculture Organization of the United Nations) (2011). The Ex Ante Carbon-balance Tool. http://www.fao.org/tc/exact/en

FAO (2012). Global Soil Partnership. http://www.fao.org/nr/water/landandwater_gsp.html

FAO/IIASA (International Institute for Applied Systems Analysis)/ISRIC-World Soil Information/ISSCAS (Institute of Soil Science, Chinese Academy of Sciences)/JRC (European Union Joint Research Centre) (2009). Harmonized World Soil Database (version 1.1). FAO, Rome, Italy and IIASA, Laxenburg, Austria. http://www.iiasa.ac.at/Research/LUC/External-World-soil-database/HTML/HWSD_Data.html?sb=4

Fearnside, P.M. and Barbosa, R.I. (1998). Soil carbon changes from conversion of forest to pasture in Brazilian Amazonia. *Forest Ecology and Management*, 108, 147-166

Fernández, R., Quiroga, A., Zoratti, C. and Noellemeyer, E. (2010). Carbon contents and respiration rates of aggregate size fractions under no-till and conventional tillage. *Soil & Tillage Research*, 109, 103-109

Foresight (2011). *The Future of Food and Farming: Challenges and choices for global sustainability*. Final Project Report. The Government Office for Science, London.

Gates Foundation (2011). 2010 *Annual Report: Strategy Refinement* (Bill & Melinda Gates Foundation). http://www.gatesfoundation.org/annualreport/2010/Pages/strategy-development.aspx

Global Soil Mapping (2011). Global Soil Information Facilities. http://www.globalsoilmap.net/category/image-galleries/global-soil-information-facilities-book

Gorham, E. (1991). Northern peatlands: role in the carbon cycle and probable responses to climatic warming. *Ecologica Applications*, 1, 182-195

Hillier, J., Walter, C., Malin, D., Garcia-Suarez, T., Mila-i-Canals, L. and Smith, P. (2011). A farm-focused calculator for emissions from crop and livestock production. *Environmental Modelling and Software*, 26, 1070-1078

Houghton, R.A. (1995). Changes in the storage of terrestrial carbon since 1850. In Lal, R., Kimble, J., Levine, E. and Stewart, B.A. (eds.), *Soils and Global Change*. Lewis Publishers, Boca Raton, Florida, USA

Janzen, H.H. (2006). The soil carbon dilemma: Shall we hoard it or use it? *Soil Biology and Biochemistry*, 38, 419-424

Jauhiainen, J., Hooijer, A. and Page, S. E. (2011). Carbon dioxide emissions from an Acacia plantation on peatland in Sumatra, Indonesia. *Biogeosciences Discuss.*, 8, 8269 - 8302

Joosten, H. (2009). *The Global Peatland CO_2 Picture. Peatland status and drainage associated emissions in all countries of the World*. Wetlands International, Ede, the Netherlands

Kapos, V., Ravilious, C., Leng, C., Bertzky, M., Osti, M., Clements, T. and Dickson, B. (2010). *Carbon, biodiversity and ecosystem services: Exploring co-benefits*. Cambodia. UNEP-WCMC, Cambridge, UK

Lal, R. (2010a). Managing Soils and Ecosystems for Mitigating Anthropogenic Carbon Emissions and Advancing Global Food Security. *BioScience* 60(9), 708-721

Lal, R. (2010b). Managing soils for a warming earth in a food-insecure and energy-starved world. *Journal of Plant Nutrition and Soil Science*, 173, 4-15

La Scala, Jr., N., de Figueiredo, E.B. and Panosso, A.R. (2011). On the mitigation potential associated with atmospheric CO_2 sequestration and soil carbon accumulation in major Brazilian agricultural activities. *Brazlian Journal of Biology* (accepted)

Malamoud, K., McBratney, A.B., Minasny, B. and Field, D.J. (2009). Modelling how carbon affects soil structure. *Geoderma*, 149, 19-26

Marks, E., Aflakpui, G.K.S., Nkem, J., Poch, R. M., Khouma, M., Kokou, M., Sagoe, R. and Sebastià, M.-T. (2009). Conservation of soil organic carbon, biodiversity and the provision of other ecosystem services along climatic gradients in West Africa. *Biogeosciences*, 6, 1825-1838

MEA (Millennium Ecosystem Assessment) (2005). Ecosystems and Human Well-Being: Synthesis. Island Press, Washington, D.C.

Milne, E., Sessay, M., Paustian, K., Easter, M., Batjes, N.H., Cerri, C.E.P., Kamoni, P., Gicheru, P., Oladipo, E.O., Minxia, M., Stocking, M., Hartman, M., McKeown, B., Peterson, K., Selby, D., Swan, A., Williams, S. and Lopez, P.J. (2010). Towards a standardized system for the reporting of carbon benefits in sustainable land management projects. *Grassland carbon sequestration: management, policy and economics* (Proceedings of the Workshop on the role of grassland carbon sequestration in the mitigation of climate change, Rome, April 2009). Integrated Crop Management, 11, 105-117

Montgomery, D.R. (2007). Soil erosion and agricultural sustainability. *Proceedings of the National Academy of Sciences*, 104, 13268-13272

Nachtergaele, F.O, Petri, M., Biancalani, R., Lynden, G. van, Velthuizen, H. van, and Bloise, M. (2011). Global Land Degradation Information System (GLADIS) Version 1.0. An Information database for Land Degradation Assessment at Global Level, LADA Technical report n. 17, FAO, Rome

Noellemeyer, E., Frank, F., Alvarez, C., Morazzo, G. and Quiroga, A. (2008). Carbon contents and aggregation related to soil physical and biological properties under a land-use sequence in the semiarid region of Central Argentina. *Soil and Tillage Research*, 99 (2),179-190

Oldeman, L.R, Hakkeling, R.T.A. and Sombroek W.G. (1991). World Map of the Status of Human-Induced Soil Degradation: An Explanatory Note (revised edition), UNEP and ISRIC, Wageningen, the Netherlands

ORNL (1998). Terrestrial ecosystem responses to global change: a research strategy. Environmental Sciences Division Publication No. 4821.

Page, S., Rieley, J.O. and Banks, C.J. (2010). Global and regional importance of the tropical peatland carbon pool. *Global Change Biology*, 17, 798-818

PlanetEarth (2005). *Soil – Earth's living skin*. International Year of Planet Earth, Trondheim, Norway. http://www.isric.nl/isric/webdocs/Docs/Soil_2.pdf

Powlson, D.S., Whitmore, A.P. and Goulding, K.W.T. (2011). Soil carbon sequestration to mitigate climate change: A critical re-examination to identify the true and the false. *European Journal of Soil Science*, 62, 42-55

Ravindranath, N.H. and Ostwald, M. (2008). *Carbon Inventory Methods: Handbook for greenhouse gas inventory, carbon mitigation and roundwood production projects. Advances in Global Change Research 29*. Springer Science + Business Media B.V.

Reed, M., Buckmaster, S., Moxey, A., Keenleyside, C., Fazey, I., Scott, A., Thomson, K., Thorp, S., Anderson, R., Bateman, I., Bryce, R., Christie, M., Glass, J., Hubacek, K., Quinn, C., Maffey, G., Midgely, A., Robinson, G., Stringer, L., Lowe, P. and Slee, B. (2010). *Policy Options for Sustainable Management of UK Peatlands. Scientific Review*. IUCN UK Peatland Programme Commission of Inquiry on Peatlands

Sarkkola, S. (ed.) (2008). *Greenhouse impacts of the use of peat and peatlands in Finland. Research Programme Final Report*. Ministry of Agriculture and Forestry, Helsinki

Schils, R., Kuikman, P., Liski, J., Van Oijen, M., Smith, P., Webb, J., Alm, J., Somogyi, Z., Van den Akker, J., Billett, M., Emmett, B., Evans, C., Lindner, M., Palosuo, T., Bellamy, P., Jandl, R. and Hiederer, R. (2008). *Review of existing information on the interrelations between soil and climate change* (ClimSoil). Final report. Brussels, European Commission

Schmidt, M.W.I., Torn, M.S., Abiven, S., Dittmar, T., Guggenberger, G., Janssens, I.A., Kleber, M., Kögel-Knabner, I., Lehmann, J., Manning, D.A.C., Nannipieri, P., Rasse, P.D., Weiner, S. and Trumbore, S.E. (2011). Persistence of soil organic matter as an ecosystem property. *Nature*, 478, 49-56

Smith, P., Martino, D., Cai, Z., Gwary, D., Janzen, H., Kumar, P., McCarl, B., Ogle, S., O'Mara, F., Rice, C., Scholes, B. and Sirotenko, O. (2007). Agriculture. In Metz, B., Davidson, O.R., Bosch, P.R., Dave, R., and Meyer, L.A. (eds.), Climate Change 2007: Mitigation. Contribution of Working Group III to the Fourth Assessment Report of the Intergovernmental Panel on Climate Change, Cambridge University Press, Cambridge, UK and New York, USA

Smith P., Gregory, P.J., Van Vuuren, D., Obersteiner, M., Havlik, P., Rounsevell, M., Woods, J., Stehfest, E. and Bellarby, J. (2010). Competition for land. *Philosophical Transactions of the Royal Society B*, 365, 2941-2957

Tanneberger, F., Tegetmeyer, C., Dylawerski, M., Flade, M. and Joosten, H. (2009). Commercially cut reed as a new and sustainable habitat for the globally threatened Aquatic Warbler. *Biodiversity and Conservation*, 18, 1475-1489

Tanneberger, F. and Wichtmann, W. (eds.) (2011). *Carbon credits from peatland rewetting. Climate – biodiversity – land use. Science, policy, implementation and recommendations of a pilot project in Belarus*. Schweizerbart Science Publishers, Stuttgart, Germany

Tarnocai, C., Canadell, J.G., Schuur, E.A.G., Kuhry, P., Mazhitova, G. and Zimov, S. (2009). Soil organic carbon pools in the northern circumpolar permafrost region. *Global Biogeochemical Cycles*, 23, GB2023

Taylor, L.L., Leake, J.R., Quirk, J., Hardy, K., Banwart, S.A. and Beerling, D.J. (2009). Biological weathering and the long-term carbon cycle: integrating mycorrhizal evolution and function into the current paradigm. *Geobiology*, 7, 171-191

Tiessen, H. and Stewart, J.W.B. (1983). Particle-size fractions and their use in studies of soil organic matter: 2. Cultivation effects on organic matter composition in size fractions. *Soil Science Society of America Journal*, 47, 509-514

Tuomi, M., Vanhalaa, P., Karhu, K., Fritze, H. and Liski, J. (2008). Heterotrophic soil respiration – Comparison of different models describing its temperature dependence. *Ecological Modelling*, 21,182-190

UNEP-WCMC (World Conservation Monitoring Centre) (2008). *Carbon and biodiversity: A demonstration atlas*. Kapos V., Ravilious, C., Campbell, A., Dickson, B., Gibbs, H., Hansen, M., Lysenko, I., Miles, L., Price, J., Scharlemann, J.P.W. and Trumper, K. (Eds.) UNEP-WCMC, Cambridge, UK

UNEP-WCMC (2009). Updated global carbon map. Poster presented at the UNFCCC COP, Copenhagen. [Article by Scharlemann et al. on methodology in preparation (2012)].

UNFCCC (2012). *Land Use, Land-Use Change and Forestry (LULUCF)*. http://unfccc.int/methods_and_science/lulucf/items/3060.php

US NRC (United States National Research Council), (2001). US NRC (United States National Research Council) (2001). *Basic Research Opportunities in the Earth Sciences*, National Academies Press, Washington, D.C.

Verburg, P.H., Neumann, K. and Nol, L. (2011). Challenges in using land use and land cover data for global change studies. *Global Change Biology*, 17, 974-989

Verwer, C.C. and Van der Meer, P.J. (2010). *Carbon pools in tropical peat forests – Towards a reference value for forest biomass carbon in relatively undisturbed peat swamp forests in Southeast Asia*. Alterra,Wageningen

Von Lützow, M., Kögel-Knaber, I., Ekschmitte, K., Matzner, E., Guggenberger, G., Marschner, B. and Flessa, H. (2006). Stabilization of organic matter in temperate soils: mechanisms and their relevance under different soil conditions –a review. *European Journal of Soil Science*, 57, 426-445

Wachs, T. and Thibault, M. (eds.) (2009). *Benefits of Sustainable Land Management*. WOCAT (World Overview of Conservation Approaches and Technologies), Berne, Switzerland

Watson, L. (2010). Portugal gives green light to pasture carbon farming as a recognised offset. *Australian Farm Journal*, January, 44-47

Wichtmann, W., Tanneberger, F., Wichtmann, S. and Joosten, H. (2010). Paludiculture is paludifuture. Climate, biodiversity and economic benefits from agriculture and forestry on rewetted peatland. *Peatlands International*, 1, 48-51

Wicke, B., Dornburg, V., Junginger, M. and Faaij, M. (2008). Different palm oil production systems for energy purposes and their greenhouse gas implications. *Biomass and Bioenergy*, 32, 1322-1337

World Bank (2010). Project Information Document: Kenya Agricultural Carbon Project. http://web.worldbank.org/external/projects/main?pagePK=64283627&piPK=73230&theSitePK=40941&menuPK=228424&Projectid=P107798

WRB (2006). World Reference Base for soil resources – A framework for international classification, correlation and communication. http://www.fao.org/ag/agl/agll/wrb/doc/wrb2006final.pdf

Zach, A., Tiessen, H. and Noellemeyer, E. (2006). Carbon turnover and 13C natural abundance under land-use change in the semiarid La Pampa, Argentina. *Soil Science Society of America Journal*, 70, 1541-1546

Closing and Decommissioning Nuclear Power Reactors
Another look following the Fukushima accident

Since the accident at Japan's Fukushima Daiichi nuclear power plant, nuclear power programmes in several countries have been under review. Germany has decided to end its programme entirely. Whatever other governments decide, the number of civilian nuclear power reactors being decommissioned is set to increase internationally as the first generations of these reactors reach the end of their original design lives.

There are plans to close up to 80 civilian nuclear power reactors in the next ten years. While many of these reactors are likely to have their operating licenses extended, they will eventually be decommissioned. The scale of the task ahead means that adequate national and international regulations, extensive funding, innovative technologies, and a large number of trained workers will be required.

Decommissioning has been carried out for a number of years without major radiological mishaps. Nevertheless, there is a need to ask: How safe is decommissioning? What are the implications of national nuclear shutdowns such as the one planned in Germany? Do countries have the necessary expertise and infrastructure to cope with the expected increase in the number of reactors to be decommissioned? And how will the very high and unpredictable costs of decommissioning be met?

What is nuclear decommissioning?

The term "decommissioning" refers to safe management – at the end of life – of many different types of nuclear facilities and sites. Decommissioning is carried out at power stations, fuel processing facilities, research reactors, enrichment plants, nuclear and radiological laboratories, uranium mines and uranium processing plants. Reactors that power submarines and ships (including ice breakers and aircraft carriers) must also be decommissioned. The biggest growth area for decommissioning is civilian nuclear power reactors (**Box 1**).

Decommissioning is only part of the final shutdown of a nuclear reactor, which begins with the removal of highly radioactive spent fuel and may end with the clean-up of an entire facility or site, including in some cases contaminated soil and groundwater (IAEA 2004a). Decommissioning involves the demolition of buildings and other structures, including the parts near the reactor core that may have become radioactive, as well as on-site handling of construction materials (mostly steel and concrete) and the packaging and transport of these materials for safe storage and disposal. Each decommissioning is associated with particular technical challenges and risks to human

The number of civilian nuclear power reactors being decommissioned is set to increase significantly in the coming decade. *Credit: visdia*

Authors: **Jon Samseth** (chair), **Anthony Banford, Borislava Batandjieva-Metcalf, Marie Claire Cantone, Peter Lietava, Hooman Peimani** and **Andrew Szilagyi**
Science writer: **Fred Pearce**

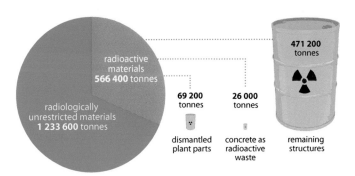

Figure 1: During the decommissioning of a nuclear power reactor large amounts of waste are generated, both radioactive (orange) and radiologically unrestricted (blue). The diagram is based on the mass flow for the decommissioned Greifswald nuclear power plant in Germany. *Source: Adapted from EWN – The Greifswald Nuclear Power Plant Site*

health and the environment. These have often been determined by choices made about reactor design and construction decades earlier (when decommissioning was little considered) as well as by operational practices over a period of years.

Most of the waste generated during decommissioning is not radiologically restricted (**Figure 1**). Radioactive decommissioning waste predominantly ranges from very low level to intermediate level radioactive (**Table 1**). High level radioactive waste (spent nuclear fuel) is generated during a reactor's operation. While the *radioactivity levels of decommissioning waste* are much lower than those of the waste generated during operations, the *volume of radioactive waste generated* during decommissioning is far greater than the volume generated during operations. Once the reactor has been closed down, radiation levels decrease over time.

State and trends in nuclear decommissioning

As of January 2012, 138 civilian nuclear power reactors had been shut down in 19 countries, including 28 in the United States, 27 in the United Kingdom, 27 in Germany, 12 in France, 9 in Japan and 5 in the Russian Federation (IAEA 2012a). Decommissioning had only been completed for 17 of them at the time of writing. Decommissioning is a complex process that takes years. The United Kingdom, for instance, completed its first decommissioning of a power reactor in 2011. This reactor, located at Sellafield, was shut down in 1981 (WNN 2011a).

The backlog of civilian nuclear power reactors that have been shut down but not yet decommissioned is expected to grow. There is also a large legacy of military and research reactors (**Box 2**). The typical design life of a civilian nuclear power reactor is 30 to 40 years. There are currently 435 such reactors in operation worldwide, with a total installed electrical capacity of 368.279 billion watts (GWe) (**Figure 3**). Of these 435 civilian nuclear power reactors, 138 are more than 30 years old and 24 are more than 40 years old (IAEA 2012a). The average age of the civilian nuclear power reactors currently in operation is 27 years (IAEA 2012a, WNA 2011a).

Many civilian nuclear power reactors will continue to operate safely beyond their original design life. Some will have their operating licences renewed for up to 60 or even 80 years (Energetics Inc. 2008). In addition, there are 63 civilian nuclear power reactors under construction with a net electrical capacity of 61 GWe (IAEA 2012a, WNA 2011b) (**Figure 4**). All nuclear reactors will have to be decommissioned some day, and the resulting radioactive waste will then need to be safely managed and disposed of (Bylkin et al. 2011).

In March 2011, a devastating 8.9 magnitude earthquake followed by a 15-metre tsunami, affected the people of Japan. Thousands of lives were lost, many people were injured and the damage to housing and infrastructure was unprecedented. The tragic earthquake and subsequent tsunami also caused the accident at the Fukushima nuclear power plant whereby radioactive material was released to the air and sea. Contamination of the reactor site

> **Box 1:** Nuclear power reactor
>
> The most common type of nuclear power reactor is the pressurized water reactor (**Figure 2**). In this type, heat generated by radioactive uranium fuel inside the reactor vessel is taken up by water and transported through a heat exchanger where steam is generated. Steam drives a turbine and generator, which produces electricity. Using a cooling source (water from a river, a lake or the sea, or from a cooling tower), the steam is condensed into water.
>
> The reactor vessel, steam generator and, in some cases, the storage pool for spent fuel (not shown in the figure) are located within a containment structure made of thick steel and/or concrete, which protects against releases of radioactivity to the environment. The parts that have become radioactive in the reactor, and require special attention during decommissioning, are the reactor vessel itself and the materials inside the vessel, including the control rods. Piping, pumps and other equipment which has been in direct contact with water that has passed through the reactor vessel or storage pool are also contaminated. A comparatively small amount of concrete may be contaminated and therefore require further treatment (O'Sullivan et al. 2010).

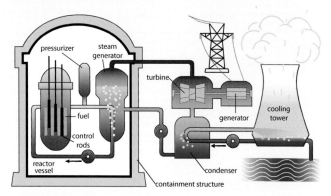

Figure 2: A pressurized water reactor produces electricity using heat generated by radioactive uranium fuel to create large amounts of steam that drive a turbine and generator. *Source: Kazimi (2003)*

Box 2: The nuclear legacy

The early years of nuclear energy left a considerable legacy of contaminated facilities, including nuclear reactors. Some are civilian in nature, but the majority are military, scientific and demonstration facilities. Until old, contaminated facilities are successfully decommissioned, they pose continuing risks and will cast a shadow over today's nuclear industry in the minds of much of the public. The challenges those involved in decommissioning must often address include incomplete facility histories and inadequate information about the state of sites and equipment. The United Kingdom's Nuclear Decommissioning Authority has reported that some facilities "do not have detailed inventories of waste, some lack reliable design drawings [and] many were one-off projects" (UK NDA 2011).

The United States Department of Energy (DOE) has undertaken to decontaminate more than 100 former research and nuclear weapons sites, covering thousands of hectares, by 2025. This will entail the management of millions of cubic metres of debris and contaminated soil, including large areas where groundwater is contaminated (Szilagyi 2012). For instance, the Oak Ridge National Laboratory in Tennessee covers 15 000 hectares with more than 100 known contaminated sites (US DOE 2011). At the larger Hanford nuclear facility in the State of Washington there are significant amounts of radioactive liquid waste (US EPA 2011a).

The DOE has successfully cleaned up complex sites such as Rocky Flats in Colorado (Tetra Tech 2012). Nevertheless, some sites may never be cleaned up for unrestricted use. In the United Kingdom, the Scottish Environment Protection Agency (SEPA) concluded in 2011 that it would do "more harm than good" to try and remove all traces of radioactive contamination from the coastline and sea bed around the Dounreay nuclear reactor site (SEPA 2011). In many countries it will be possible to reuse decommissioned sites that are not fully cleaned up for new nuclear applications (IAEA 2011a).

Reactors built to power submarines or ships are one type of legacy concern. Decommissioning a typical nuclear submarine produces more than 800 tonnes of hazardous waste (Kværner Moss Technology 1996). At the end of the Cold War there were over 400 nuclear submarines, either operational or being built, mainly in the former Soviet Union and in the United States (WNA 2011d). Many nuclear submarines have been withdrawn from service and most await decommissioning. The United States has decommissioned a number of them, with their reactors removed, properly packaged and staged for disposal at Hanford. Before 1988, some 16 reactors from dismantled nuclear submarines in the former Soviet Union were disposed of by dumping at sea (Mount et al. 1994, IAEA 1999).

---

and its surroundings made an area with a radius of about 30 km uninhabitable or unsuitable for food production – in some cases for months or years to come. Japan's power-generating capacity has been seriously affected, and the political impact in other countries has led some governments to question their reliance on nuclear energy. So far, only Germany has decided to end nuclear power generation (BMU 2011, WNA 2011c). However, the debate continues in a number of other countries (Okyar 2011). The company which built many of Germany's nuclear reactors has announced that it does not plan to build any more reactors anywhere in the world (Der Spiegel 2011). As some civilian nuclear power reactors that had previously been expected to operate for many more years join those nearing the end of their design life, the total number awaiting decommissioning is likely to increase significantly.

Three approaches to decommissioning

There are three generally accepted approaches to decommissioning: immediate dismantling, deferred dismantling and entombment. Each approach requires early and clear decisions about the timing of the closure of facilities and intended future use of the site (**Figure 5**). Each also requires adequate funding, trained personnel, regulatory oversight and waste storage and disposal facilities (IAEA 2006).

Immediate dismantling: All equipment, structures and other parts of a facility that contain radioactive contaminants are removed (or fully decontaminated) so that the site can be treated as uncontaminated

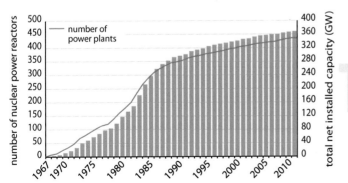

Figure 3: By early 2012 the number of nuclear power reactors in the world had increased to 435. Total installed electrical capacity has increased relatively more rapidly than the number of reactors. *Source: IAEA (2012)*

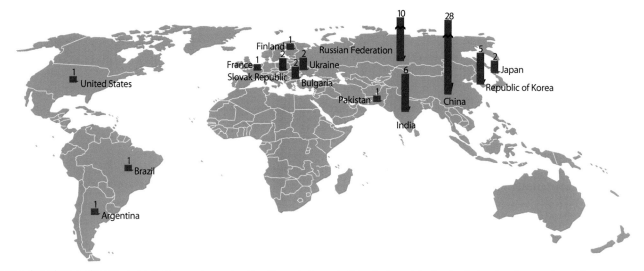

Figure 4: Sixty-three nuclear power reactors are under construction. The majority are in China, India and the Russian Federation. *Source: Adapted from IAEA (2012)*

for either unrestricted or more restricted use (sometimes referred to as a "greenfield" site). This internationally agreed approach has the advantage that experienced operational staff from the facility are still available who know the history of the site, including any incidents in the past that could complicate the decommissioning process. Immediate dismantling also avoids the unpredictable effects of corrosion or other degradation of the reactor parts over an extended period, eliminates the risk of future exposure to radiation, and removes a potential blight on the landscape. A disadvantage of this approach is that levels of radioactivity in the reactor parts are higher than in the case of deferred dismantling. This means that greater precautions must be taken during dismantling, and that larger volumes of decommissioning waste will be classified as radioactive.

Deferred dismantling: After all the spent fuel is removed, the plumbing is drained, and the facility is made safe while dismantling is left for later. This approach is often called "safe enclosure". The deferral periods considered have ranged from 10 to 80 years (Deloitte 2006). For instance, the Dodewaard reactor in the Netherlands was shut down in 1997 but will not be decommissioned until at least 2047 (IAEA 2004b). Deferred decommissioning has the advantage of allowing radioactive materials to decay to lower levels of radioactivity than in the case of immediate dismantling (**Box 3**). This reduces both disposal problems and risks of harm to workers. In the meantime, robotic and other types of techniques that make dismantling safer and cheaper may undergo further development. A disadvantage is that some materials, including concrete and steel, may deteriorate, making the eventual decommissioning more difficult. Moreover, personal knowledge of a site's history will be lost as time passes.

Entombment: Once the spent fuel has been removed, reactors can be entombed. This involves encasing the structure in highly durable material such as concrete while its radioactivity decays. Entombment is a relatively new approach that is mainly considered in special cases (examples are small research reactors or reactors in remote locations). It can reduce worker exposure to radioactivity since there is less handling of contaminated materials. However, long-term maintenance and monitoring are

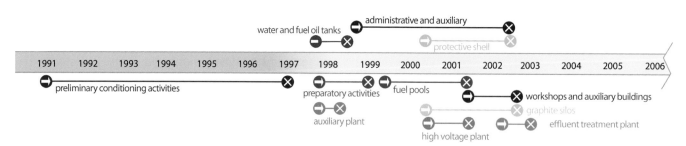

38 UNEP YEAR BOOK 2012

 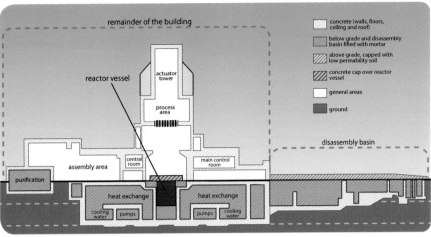

Nuclear power plant prior to decommissioning | Cross-section of the entombment structure

Figure 6: Entombment at the Savannah River site, United States. All the spent fuel and other high level waste was removed from the reactor, as well as that portion of waste/contamination indicated as unacceptable based on rigorous risk and performance assessments. Entombment entailed subsequent filling with specialized mortar of all subsurface spaces where contamination existed. Above-ground uncontaminated areas were generally left as they were. To provide additional protection against water intrusion and infiltration, the building was left standing and will be monitored over the long term. *Source: Adapted from US DOE (2012)*

required. Five reactors have been entombed in the United States, with the entombment of two reactors at the Savannah River site completed in 2011 (**Figure 6**).

The challenges of decommissioning

Decommissioning has been accomplished so far without creating significant additional health and safety or environmental risks, although it has occasionally revealed unsuspected past contamination from nuclear operations (WNA 2011a). An adequate legal framework nevertheless needs to be in place, with clear responsibilities assigned to different actors including regulatory authorities. Otherwise, risks could increase as the number of decommissionings increases; as pressures grow in some countries to speed up the closing and decommissioning of nuclear power plants, shorten overall schedules and cut costs; and as decommissioning begins in countries with little or no previous experience and insufficient waste management capacity. More experience should eventually contribute to improved techniques and reduced costs. However, unless the accelerated phase-out of nuclear reactors is carefully managed, with adequate regulatory oversight, it could lead to overly hasty decisions to decommission or to reactors standing idle for many years before decommissioning finally takes place. The latter situation, if not properly monitored and managed, could lead to increased risks of releases of radioactive contaminants to the environment and exposure of nearby populations (IAEA 2007).

Smarter dismantling

A critical aspect of decommissioning is that dismantling needs to be carried out in such a way that radioactive and non-radioactive materials are separated. This minimizes the amount of waste that will require special treatment because of its radioactivity. Separation also maximizes the amount of materials such as steel and aluminium that can be recycled, as well as the amount of concrete rubble that can be reused on site (Dounreay 2012). Some materials may need to be dismantled and decontaminated on-site. The complex task of dismantling requires good information at the beginning of the process about the radiological characteristics and state of the reactor, including its operational history, such as incidents and accidents, and the presence of any spent fuel debris.

The need to dismantle structures whose purpose has been to protect workers during the reactor's operation can make decommissioning more difficult. For instance, steel pipes

Figure 5: The dismantling and decommissioning of the Vandellós I civilian nuclear power reactor in Spain is taking place in three main phases: reactor shutdown and preliminary activities (1991-1997); removal of non-reactor structures (1998-2003); and dismantling of the reactor vessel (around 2028). The third phase is scheduled to begin after a 25-year dormancy period, during which the reactor is to remain under close surveillance. *Source: Adapted from ENRESA (2009)*

> **Box 3:** Radiation associated with decommissioning

The bulk of the radioactive waste from decommissioning consists of very low level and low level waste, mostly steel and concrete. Higher level radioactive waste from decommissioning consists mainly of reactor components. This waste contains isotopes that emit radiation as they decay. The initial release of radiation decreases rapidly due to the relatively short half-life of a number of isotopes. After 50 years, the radiation level in most decommissioning waste decays to a small percentage of the initial level.

Isotope	Half-life (years)
C-14	5 730
Ni-59	75 000
Ni-63	96
Fe-55	2.7

At very high doses radiation can cause radiation sickness, cancers and even near-term or immediate death, as in the case of on-site workers at the time of the Chernobyl accident. At lower doses it may induce cancers and genetic damage. At doses normally received during operations or decommissioning, however, risks to workers should be negligible.

The radiation encountered during decommissioning and the disposal of the waste generated is almost exclusively beta and gamma radiation (**Figure 7**). Decommissioning risks are mostly associated with exposure to these types of radiation. Since the waste from decommissioning is most commonly in solid form, only unintended releases of radioactive dust generated during demolition has the potential to result in exposure of the general public (US EPA 2011b).

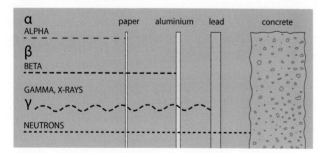

Figure 7: Alpha, beta, gamma and neutron radiation differ in their ability to penetrate materials. Alpha particles do not penetrate far. They can be stopped by a sheet of paper, while beta particles can be stopped by a thin piece of aluminium, gamma rays by heavy metals, such as lead, and neutrons by concrete or water. *Source: WNA (2011e)*

carrying highly radioactive liquids are often encased in concrete. This makes decommissioning more complex, in that the pipes may be radioactive while the large volumes of concrete in which they are embedded are not. The contaminated material will either have to be removed separately or segregated later (O'Sullivan et al. 2010).

A key to reducing the volume of contaminated waste is to improve the separation of materials during decommissioning. But reconciling this practice with the minimization of worker exposure may be difficult. Evaluations are therefore carried out prior to decommissioning in order to choose appropriate approaches that make use of manual or remote control techniques. In many cases remotely operated vehicles, manipulator arms and robots can be used to cut waste materials into smaller pieces. Further development of such technologies will be invaluable, as they can reduce volumes of radioactive waste through more selective cutting, thus reducing both costs and radiological risks.

Experience with decommissioning the first generations of nuclear reactors suggests that decommissioning would have been easier and less expensive if they had been designed with this stage in mind (OECD/NEA 2010a). Few old reactors incorporate design features that help or simplify decommissioning. Nuclear power plants currently in operation commonly have a decommissioning plan, as preliminary plans are often a requirement for the application for a licence to operate a nuclear facility (OECD/NEA 2010a). Decommissioning plans should be updated regularly, with a detailed scheme drawn up at least two years before the scheduled shutdown (IAEA 2008, 2011b). However, some

Compactable low level waste may include radioactive clothing, glass and building materials. *Credit: Sellafield Ltd.*

Cut through a barrel containing intermediate level liquid waste solidified in concrete. Intermediate level waste consists of heavily contaminated materials, such as fuel rod casings or decommissioned parts of the reactor vessel. This waste requires radiation shielding. Storage time will depend upon which radioactive isotopes are present in the waste. Radioactive liquids are solidified before disposal. *Credit: Dounreay*

reactors are inevitably shut down early because of a change of policy, an accident or a natural disaster (**Box 4**).

Resources and capacity

Several countries have developed expertise in decommissioning. In the United States, for instance, 1 450 government nuclear facilities of various kinds have been fully decommissioned, including a number of reactors (US DOE 2012). While such expertise in some countries is ground for optimism, a number of other countries have yet to develop expertise and infrastructure on the scale that will be necessary in the future. Universities and technical centres in a number of countries are setting up training programmes or undertaking research and development specifically related to decommissioning. Much of this activity is focused on automatic equipment and innovative methods of working in a radioactive environment.

Future decommissioning of civilian nuclear power reactors will compete for expertise, resources and waste disposal facilities with the decommissioning of many military and research reactors and other facilities. More than 300 such reactors, both small and large, have been taken out of operation (WNA 2011a), but the majority have not yet been decommissioned.

Public acceptability

Public acceptability is critical to the future of nuclear power (OECD/NEA 2010b). Whether nuclear power plants are decommissioned immediately or after some delay, what happens to radioactive waste, and whether the end result is a greenfield site, entombment or something in between can depend on acceptance by the public as

Box 4: Managing damaged reactors

Decommissioning requires a safety assessment to be approved by regulatory authorities, and both an environmental impact assessment (EIA) and an environmental impact statement (EIS) to be completed. Decommissioning in the aftermath of a major accident such as Three Mile Island (the United States), Chernobyl (Ukraine) or Fukushima (Japan) is quite different from planned decommissioning at the end of a facility's lifetime. Different types of planning, equipment and funding are needed. A damaged reactor may contain exposed nuclear fuel and its containment may be compromised. The reactor and associated facilities must be stabilized and made safe before dismantling or entombment take place.

In 1979 the Three Mile Island No. 2 reactor experienced a partial meltdown during which the core overheated. The operators carried out a clean-up, removing fuel, decontaminating radioactive water and shipping radioactive waste to a disposal site. Fuel and debris from the molten core were moved to a government facility, where they are now in dry storage awaiting a decision on the final disposal location. The reactor itself is in "monitored storage" until the No. 1 reactor is shut down. Both reactors will then be decommissioned (US NRC 2009).

In 1986 the Chernobyl No. 4 reactor exploded and burned, releasing large amounts of radioactive material to the air. The fire caused by the explosion was extinguished after several hours, but the graphite in the reactor burned for several days. It took half a year to encase the reactor in a concrete sarcophagus. This will not be the final entombment, however. The sarcophagus has deteriorated to such an extent that water is leaking in and it may be collapsing. There are plans to put a new containment around the sarcophagus by the end of 2015, so that the decaying structure and the fuel and other contaminated material inside can be removed safely to a new waste store (Wood 2007, Yanukovych 2011).

In December 2011 the Tokyo Electric Power Company (Tepco), the Ministry of Economy, Trade and Industry's Agency for Natural Resources and Energy, and the Nuclear and Industrial Safety Agency of Japan announced the first roadmap for decommissioning of the Fukushima reactors. It calls for the removal of fuel remaining in the storage pools within ten years. Starting in ten years, the fuel that constituted the cores of the reactors will be removed. This will be a very complex task, as the extent of damage to the cores is unknown. One of the reactor cores is thought to have melted through the reactor vessel and into the concrete floor below the reactor. To remove the cores will take another 10-15 years. Final demolition of the reactor structures will to be completed in 30-40 years (WNN 2011b).

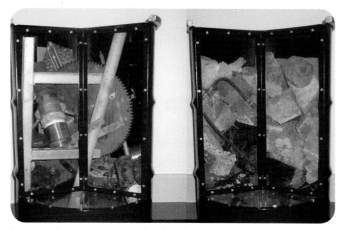
Interim packaging and storage of radioactive waste. *Source: Nuclear Decommissioning Authority, United Kingdom*

much as on technical considerations. Intense decommissioning activity may be disliked by neighbours, but it can remove a blight on the landscape and allow new land use. Entombment, on the other hand, is not only visually unattractive, but maintaining a reactor in "safe mode" requires permanent security arrangements (OECD/NEA 2010b).

Some operators fear public debate, while others embrace it. The Nuclear Decommissioning Authority in the United Kingdom, for instance, is taking a more open approach than in the past (UK DTI 2002). Increased openness can have demonstrable success. In the United States, the National Aeronautics and Space Administration (NASA), which operates the Plum Brook research reactor in the State of Ohio, responded to public concern about decommissioning with a programme of community workshops, websites, videos, reactor media tours and open days. This potentially controversial decommissioning eventually gained local support (IAEA 2009a). The Forum of Stakeholder Confidence, created in 2000 by the intergovernmental Nuclear Energy Agency (NEA), facilitates sharing of experience in addressing the societal dimension of radioactive waste management. This body explores ways to maintain a constructive dialogue with the public in order to strengthen confidence in decision-making processes, which may involve players at the national, regional and local levels (OECD/NEA 2011).

Unpredictability of decommissioning requirements

Decisions resulting from countries' reappraisal of their nuclear power programmes following the Fukushima accident will have important implications for their national decommissioning programmes. They will also raise questions about whether the necessary skills, expertise, funding and infrastructure are in place to meet new and unanticipated decommissioning demands.

Of Japan's 50 remaining nuclear power reactors, only 5 are operating at the time of writing (IAEA 2012a, WNN 2012a). Any of these reactors could eventually be restarted once stress tests are performed, improved protection against tsunamis is in place, and approval from both the government and local authorities has been obtained. The government closed the Hamaoka nuclear power plant temporarily in 2011 because of fears concerning a future large earthquake in its area. This plant will be reopened when better protection against tsunamis has been provided (WNN 2011d).

Germany's decision to phase out all of its nuclear power plants by 2022 means bringing forward the closure of 13 currently operating plants (WNA 2011d). These plants' early phase-out will be costly. It will also require safe handling of very large volumes of decommissioning waste or, if decommissioning is deferred, the safe maintenance of a number of mothballed facilities. Considerable demands will be made on Germany's decommissioning expertise and infrastructure.

Those involved in decommissioning in any country need to be prepared for the unexpected. For instance, legislators, regulators or lawyers could intervene to initiate or halt decommissioning. In 2010 the Vermont State Senate in the United States revoked the license of the Vermont Yankee nuclear power plant because of concerns about leaks of radioactive tritium gas, as well as allegations that misleading statements on this issue had been made by the operators. The plant was scheduled to close in March 2012, but the operators were successful in their legal challenge to the state's right to shut it down (WNN 2011d, 2012b).

Costs and financing of decommissioning

The costs of decommissioning nuclear power reactors vary greatly, depending on the reactor type and size, its location, the proximity and availability of waste disposal facilities, the intended future use of the site, and the condition of both the reactor and the site at the time of decommissioning. Methods for carrying out cost estimates have been developed (OECD/NEA 2010c). However, published data on the costs of the small number of decommissionings completed so far are sparse (OECD/NEA 2010c, US GAO 2010). Estimates of future costs vary hugely.

Decommissioning costs represent a substantial share of the costs of a nuclear power reactor's operation (**Figure 8**). On the other hand, they may represent only a small percentage of the income generated by a civilian nuclear power reactor over a 40-year period. In the United States, the average costs of decommissioning a nuclear power reactor have been around US$500 million or approximately 10-15 per cent of the initial capital cost. In France, it is estimated that decommissioning the small Brennilis reactor (in operation from 1967 to 1985) will equal 59 per cent of the reactor's

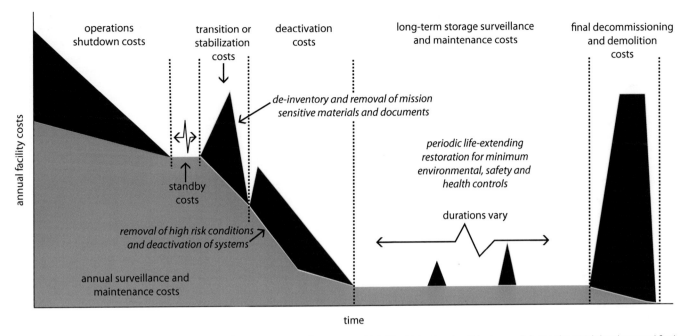

Figure 8: Decommissioning a nuclear power plant takes many years and costs vary widely. The highest costs will be incurred during the initial shutdown and final decommissioning and demolition. Any intervening period of standing by will be less expensive. These factors may influence decisions on how rapidly decommissioning will take place. *Source: United States Department of Energy (2010)*

initial cost. This estimate rose by 26 per cent between 2001 and 2008, to almost €500 million – as much as 20 times the original estimate (Cour des comptes 2005, 2012). In the United Kingdom, the government's financial provision for decommissioning rose from £2 million in 1970 to £9.5 billion in 1990 and £53.7 billion in 2011 (Huhne 2011). It is clear that decommissioning can sometimes be much more expensive than originally budgeted (OECD/NEA 2010d). As more experience is gained, this type of uncertainty should diminish and the costs come down.

In many countries the responsibility for funding decommissioning activities rests with the owner, in compliance with the polluter pays principle (Deloitte 2006, Wuppertal 2007). Nevertheless, governments are responsible for ensuring that adequate funds are generated during the operation of nuclear power plants on their territory to pay these high and sometimes unpredictable costs. The extent to which funds are protected against financial crises is not always clear. Investment funds will not necessarily deliver the anticipated returns. In any event, governments are likely to be the funders of last resort (SwissInfo 2011).

In 2006 the European Commission issued a recommendation and a guide on the management of financial resources for the decommissioning of nuclear installations and the handling of spent fuel and radioactive waste (EU 2006a, b). Furthermore, under a recent EU Directive establishing a Community framework for the responsible and safe management of spent fuel and radioactive waste, all Member States are to ensure that funding resources are available for decommissioning (EU 2011). Many

---

Box 5: Regulating decommissioning at the global level

The Joint Convention on the Safety of Spent Fuel Management and on the Safety of Radioactive Waste Management is the first legal instrument to directly address, among other issues, the management of radioactive waste from decommissioning on a global scale (IAEA 2011c). The Joint Convention, which entered into force on 18 June 2001, has been ratified by 62 countries. Its Article 26 specifies that "Each Contracting Party shall take the appropriate steps to ensure the safety of decommissioning of a nuclear facility. Such steps shall ensure that: (i) qualified staff and adequate financial resources are available; (ii) the provisions of Article 24 with respect to operational radiation protection, discharges and unplanned and uncontrolled releases are applied; (iii) the provisions of Article 25 with respect to emergency preparedness are applied; and (iv) records of information important to decommissioning are kept."

Table 1: Radioactive waste classification. *Source: Adapted from IAEA (2009b)*

	very low level waste (VLLW)	low level waste (LLW)	intermediate level waste (ILW)	high level waste (HLW)
radioactivity	contains very limited concentrations of long-lived radioactive isotopes with activity concentrations usually above the clearance levels	contains limited concentrations of long-lived radioactive isotopes but has high radioactivity	contains long-lived radioactive isotopes that will not decay to a level of activity concentration acceptable for near surface disposal	contains levels of activity concentration high enough to generate significant quantities of heat by radioactive decay or with large amounts of long-lived radioactive isotopes
examples of waste sources	concrete rubble, soil	clothing, glass, building materials	fuel rod casings, reactor vessel part	debris of spent fuel
isolation	engineered surface landfill	near surface disposal at depth up to 30 metres	shallow disposal at depth from a few tens to a few hundred metres	deep geological formations
need shielding	no	no	yes	yes
need cooling	no	no	no	yes

European governments – but not all – have ensured that such funding is available. The funding systems vary. In Spain, for instance, a public company is in charge of funding, while in Slovakia this is the responsibility of the Ministry of Economy. At the global level, the need to have adequate resources available for decommissioning is being addressed by the Joint Convention on the Safety of Spent Fuel Management and on the Safety of Radioactive Waste Management (**Box 5**).

Risks associated with decommissioning

The risks of large-scale releases of radioactivity during decommissioning are much lower than during a reactor's operations. Once the nuclear fuel has been removed, most of the radioactivity is gone. When the tanks and plumbing are drained, the majority of the radioactive materials that remain are in solid form, which is easier to handle and less likely to enter the environment. However, the non-routine and hands-on nature of the work means risks related to worker exposure are higher during decommissioning than during operations.

Types and quantity of radioactive waste

During operations, a nuclear reactor produces isotopes that give out potentially harmful radiation as they decay. Their half-life (the time it takes to halve the radioactivity of the isotope) varies from seconds to millions of years. Those with a half-life of more than ten days may contribute to radioactive waste. The waste needs to be kept safe until the process of decay reduces the radioactivity levels of the materials. For storage and disposal, it is usually classified into different types (very low level, low level, intermediate level and high level radioactive waste) according to risks and decay time (**Table 1**).

Most of the high level radioactive material that finally contributes to high level radioactive waste is the spent fuel regularly removed from operating reactors. A typical 1000-MW reactor produces about 27 tonnes of this waste per year (WNA 2011e). The amount of spent fuel produced by the world's reactors is barely enough to fill two Olympic size swimming pools every year. Although the volumes are relatively small, high level waste contains 95 per cent of the radioactivity in waste from the nuclear power industry. It will need to be kept isolated for thousands of years.

A typical disposal method for low level radioactive waste is burial underground. Care needs to be taken that water does not transport radioactive isotopes beyond the burial site. *Credit: US NRC*

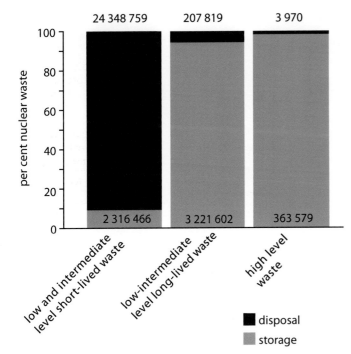

Dismantling a 1000-MW reactor generates around 10 000 m³ of VLLW, LLW and ILW, but that amount may be greatly reduced with proper management and use of robots to more selectively separate the more radioactive parts from the rest (McCombie 2010). This waste can include large amounts of construction materials, along with steel reactor vessel equipment, chemical sludges, control rods, and other types of material that have been in close proximity to reactor fuel. The radioactivity of the waste generated during decommissioning will usually be negligible within a few decades. Nevertheless, this waste requires safe handling, storage and disposal until that time.

Of the low and intermediate level long-lived radioactive waste produced during decommissioning, only 7 per cent has been disposed of so far (**Figure 9**). The remaining 93 per cent remains in storage and is awaiting safe disposal. Many countries have established radioactive waste management agencies, but there is a long way to go before these agencies are equipped to handle the volumes of waste likely to emerge from future decommissioning (CoRWM 2006). Disposal facilities for very low level waste already exist in countries producing nuclear power.

Figure 9: Decommissioning generates waste that can be categorized as low, intermediate and high level nuclear waste. The total waste inventory shows the percentage of nuclear waste by type in storage, compared with that sent to disposal. Volumes are expressed in cubic metres and based on data reported by countries using the older 1994 IAEA waste classification, according to which low level waste and intermediate level waste were combined into two subgroups: short-lived and long-lived. Very low level waste was not distinguished as a separate category. *Source: Adapted from IAEA (2011d)*

According to current waste management practices, high level waste will ultimately require disposal in deep geological formations. While some countries, including Finland, France and Sweden, have selected sites, no country yet has an operational high level radioactive waste disposal facility. This is partly related to costs, partly to public opposition to proposed sites (WNA 2011f), and partly to the fact that insufficient time has elapsed for the spent fuel and other high level radioactive waste to become cool enough to be placed in a permanent repository. In the first 20 to 30 years after final shutdown, part of the inner components to be handled by decommissioning belongs to the high level waste class.

After the spent fuel is removed, decommissioning produces only small amounts of high level waste (HLW), most of which is nuclear fuel debris left behind after the last fuel was removed from the reactor. However, decommissioning typically generates two-thirds of all the very low, low and intermediate level waste (VLLW, LLW and ILW) produced during a reactor's lifetime.

Potential pathways for exposure to radioactivity

Decommissioning activities such as the cutting up of equipment have the potential to disperse radioactive dust or gas (Shimada et al. 2010) (**Figure 10**). Such air emissions present risks primarily to workers. These emissions need to be contained or ventilated

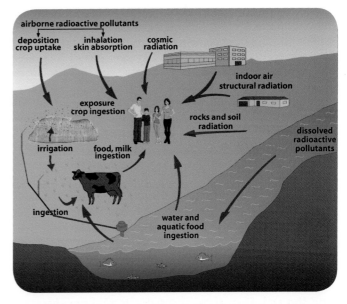

Figure 10: Radiation exposure pathways. During decommissioning, airborne radioactive dust particles may be released unintentionally if a mishap occurs. *Source: Adapted from Arizona State University (2011)*

safely, using filters to catch the dust. Highly contaminated reactor components can sometimes be cut up under water. This provides shielding for workers and prevents radioactive releases to the air. Waste stored on-site poses potential risks if the storage equipment suffers corrosion or dissolution, or in case of fire. There are also risks related to fires or floods at decommissioning sites that release radioactive materials to the air, soil or groundwater (for instance, from areas where waste is processed or stored). If water penetrates the disposal site, it can dissolve radioactive isotopes and transport them to the water system. However, most isotopes encountered during decommissioning are relatively insoluble or have a short half-life.

The potential for large-scale releases of radioactivity beyond a nuclear power plant during decommissioning is much less than that during its operation. However, low level releases can occur over short distances via the air or surface and groundwater. Careful planning, and the use of barriers and local and perimeter monitoring, can help protect against such releases.

Unanticipated conditions may be discovered during the decommissioning of a facility that has been in operation for several decades. There may be unexpected spent fuel debris within the reactor, although this occurs more often in research reactors and other reactors not used for power generation. Radioactive contamination beneath the reactor site that has not yet migrated to the underlying groundwater may not be detected until the facility has been demolished. Although this case represents the exception rather than the rule, when the Yankee nuclear power plant in the State of Connecticut (United States) was dismantled (**Figure 11**), decommissioners discovered 33 000 m³ of radioactively contaminated soil that had to be removed and disposed of, greatly adding to the cost of making the site safe (EPRI 2008). Decommissioning itself may, through excavations or other activities, increase the risk of radioactive contamination migrating from soils to surface or groundwater.

During operations, parts of a nuclear power plant near the reactor core become radioactive. To keep the doses of radiation received by workers during decommissioning as low as reasonably achievable – and below regulatory limits – there is a need for extensive work planning, administrative and physical controls, use of protective clothing, and a comprehensive monitoring programme. Doses can be further reduced through the use of robots and other remote techniques that enable removal of workers from locations near radioactive hazards. To date, the level of exposure during decommissioning has been below regulatory limits.

Figure 11: The Connecticut Yankee nuclear power plant was successfully decommissioned and the site restored to a greenfield. The pictures show progress over time at the start (June 2003), during operations (January 2006) and after decommissioning was completed (September 2007).
Credit: Connecticut Yankee Atomic Power Company

Nuclear power plants damaged as a result of accidents, such as those at Chernobyl and Fukushima, must be handled very differently from plants at the end of their expected design life. Contaminated material may have been released over long distances, in which case emergency responses will be required to prevent further releases. Once radioactive releases have been halted and the damaged plant has been stabilized, the nuclear fuel has to be removed from the reactor, which could be damaged. Only then can work begin to decommission the facility and clean up the site and surrounding areas.

Historically, discussions of the environmental impacts of nuclear activities (including decommissioning) have focused almost exclusively on human health risks. In 1991, the International Commission on Radiological Protection (ICRP) gave its opinion that "the standard of environmental control needed to protect man to the degree currently thought desirable will ensure that other species are not put at risk." The Commission currently indicates that this view was too narrow. Instead, it states that risks to biodiversity and ecosystems from decommissioning and other activities cannot be assumed from those calculated for humans (Higley et al. 2004).

Since 2007 the ICRP has been developing radiation dose reference levels for 12 animals and plants, from duck to deer and from seaweed to earthworms (ICRP 2007). The reference levels are not regarded as limits, but as thresholds for further consideration (Andersson et al. 2009). Rather than eliminating all risks to individual organisms, the aim has been to "prevent or reduce the frequency of deleterious radiation effects to a level where they would have a negligible impact on the maintenance of biological diversity, the conservation of species, or the health and status of natural habitats" (ICRP 2007).

Lessons learned

Decommissioning is not simply demolition. It is the systematic deconstruction of a contaminated, complex nuclear facility made up of a reactor with many large components such as the reactor vessel, steam generators, pumps and tanks, and supporting systems including thousands of metres of pipes – along with even greater volumes of construction materials. This type of deconstruction requires considerable time and funding, detailed planning and precise execution, on a level similar to that required in order to build a nuclear facility. It also requires a similar degree of expertise and regulatory control.

While decommissioning is still a maturing industry in different parts of the world, it is fast-growing. There are considerable geographical differences in degrees of expertise. A few countries have decades of experience. For others, such experience is all in the future. Important knowledge has been gained, but the lessons learned are not yet reflected in standard practice internationally. The International Atomic Energy Agency (IAEA) has established an international decommissioning network to facilitate exchanges of experience among countries (IAEA 2012b).

Ensuring that these important lessons are applied globally in time for the anticipated boom in decommissioning is of critical importance. International agencies and the owners and operators of nuclear facilities, in particular, need access to all the information available from contractors. There is a case for international and national laws that would require the sharing of such information. This would include expertise obtained when things have gone wrong, as it is often then that the most important lessons can be learned. There is a strong need to keep considerations of commercial confidentiality from getting in the way.

The nuclear industry will need to continue to innovate and develop new approaches and technologies that facilitate a "smarter" decommissioning process, meaning one that is safer, faster and cheaper. Additionally, meeting the decommissioning challenge will require policies and measures that support the continuing evolution of these decommissioning improvements. Research could further contribute to building the knowledge foundation and provide a strong scientific underpinning for decommissioning.

The coming decade will probably witness the rapid expansion of decommissioning activity, costing tens of billions of dollars. The decommissioning industry's performance will be critical to the future of nuclear power generation. The challenges are technical, but also political, financial, social and environmental.

Experience shows that decommissioning can be carried out in a safe, timely and cost-effective manner. One lesson emerging is that nuclear power plants should be designed, from the beginning, for safe and efficient decommissioning as well as for their safe operation, accident prevention, and safety with respect to the potentially affected public and the environment. The first generations of nuclear power plants were designed with little thought for decommissioning, resulting in costs that might otherwise have been avoided. Today many operators and regulatory agencies incorporate features that will help or simplify decommissioning in the design of new nuclear power plants.

References

Andersson, P., Garnier-Laplace, J., Beresford, N.A., Copplestone, D., Howard, B.J., Howe, P., Oughton, D. and Whitehouse, P. (2009). Protection of the environment from ionizing radiation in a regulatory contex (PROTECT): proposed numerical benchmark values. *Journal of Environmental Radioactivity*, 100, 1100-1108

Arizona State University (2011). Radiation Exposure Pathways. http://holbert.faculty.asu.edu/eee460/pathways.jpg

BMU (Bundesministerium für Umwelt, Naturschutz and Reaktorsicherheit; Federal Ministry for Environment, Nature Conservation and Nuclear Safety) (2011). Questions and answers about transforming our energy system. http://www.bmu.de/english/energy_efficiency/doc/47609.php

Bylkin, B.K., Pereguda, V.I., Shaposjnikov, V.A. and Tikhonovskii, V.L. (2011). Composition and structure of simulation models for evaluating decommissioning costs for nuclear power plant units. *Atomic Energy*, 110(2), 77-81

CoRWM (Committee on Radioactive Waste Management) (2006). *Managing our Radioactive Waste Safely. CoRWM recommendations to Government*. DoRWM Doc 700. July. http://news.bbc.co.uk/nol/shared/bsp/hi/pdfs/310706_corwmfullreport.pdf

Cour des comptes (2005). *Le démantèlement des installations nucléaires et la gestion des déchets radioactifs : Rapport au Président de la République suivi des réponses des administrations et des organismes intéressés*. Paris, France. http://www.ccomptes.fr/fr/CC/documents/RPT/RapportRadioactifsnucleaire.pdf

Cour des comptes (2012). *Les coûts de la filière d'électronucléaire. Rapport public thématique*. Janvier 2012. Paris, France. http://www.ccomptes.fr/fr/CC/documents/RPT/Rapport_thematique_filiere_electro-nucleaire.pdf

Deloitte (2006). Nuclear Decommissioning and Waste: A Global Overview of Strategies and the Implications for the Future. *Deloitte Energy and Resources*, May. http://deloitte-ftp.fr/Lot-B-Energie-ressources/doc/NuclearDecommissioning.Mai06.pdf

Der Spiegel (2011). Response to Fukushima: Siemens to Exit Nuclear Energy Business, 19 September. http://www.spiegel.de/international/business/0,1518,787020,00.html

Dounreay (2012). Non-radioactive waste. Waste is a product of decommissioning. Dounreay Site Restoration Ltd. http://www.dounreay.com/waste/nonradioactive-waste

ENRESA (Empresa Nacional de Residuos Radiactivos, S.A.) (2009). The dismantling and decommissioning of the Vandellós I nuclear power plant step by step. http://www.enresa.es/activities_and_projects/dismantling_and_decommissioning/dismantling_vandellos_i_step_by_step

EPRI (Electric Power Research Institute) (2008). *Power Reactor Decommissioning Experience*. Technical Report ID 1023456. Palo Alto, California, USA. http://my.epri.com

EU (European Union) (2006a). Commission Recommendation of 24 October 2006 on the management of financial resources for the decommissioning of nuclear installations, spent fuel and radioactive waste (2006/851/Euratom). http://eurlex.europa.eu/LexUriServ/LexUriServ.do?uri=OJ:L:2006:330:0031:0035:EN:PDF

EU (2006b). Guide to the Commission Recommendation on the management of financial resources for the decommissioning of nuclear installations (2006/851/Euratom). http://ec.europa.eu/energy/nuclear/decommissioning/doc/2010_guide_decommissioning.pdf

EU (2011). Council Directive 2011/70/Euratom of 19 July 2011 establishing a Community framework for the responsible and safe management of spent fuel and radioactive waste. http://eur-lex.europa.eu/LexUriServ/LexUriServ.do?uri=OJ:L:2011:199:0048:0056:EN:PDF

EWN (Energiewerke Nord GmbH) (1999). EWN – *The Greifswald Nuclear Power Plant Site*. http://ec-cnd.net/eudecom/EWN-WasteManagement.pdf

Higley, K.A., Alexakhin, R.M. and McDonald, J.C. (2004). Dose limits for man do not adequately protect the ecosystem. *Radiation Protection Dosimetry*, 109(3), 257-264

Huhne, C. (2011). Chris Huhne Speech to the Royal Society: Why the future of nuclear power will be different; speech by the Secretary of State for Energy and Climate Change, 13 October. http://www.decc.gov.uk/en/content/cms/news/ch_sp_royal/ch_sp_royal.aspx

IAEA (International Atomic Energy Agency) (1999). *Radioactivity in the Arctic Seas. A Report of International Arctic Seas Assessment Project (IASAP)*. IAEA TECDOC 1075. http://www-pub.iaea.org/MTCD/publications/PDF/te_1075_prn.pdf

IAEA (2004a). Status of the decommissioning of nuclear facilities around the world. http://www-pub.iaea.org/MTCD/publications/PDF/Pub1201_web.pdf

IAEA (2004b). *Transition from Operation to Decommissioning of Nuclear Installations*. http://www-pub.iaea.org/MTCD/publications/PDF/TRS420_web.pdf

IAEA (2006). *IAEA Safety Standards. Decommissioning of Facilities Using Radioactive Material*. Safety Requirements WS-R-5. http://www-pub.iaea.org/MTCD/publications/PDF/Pub1274_web.pdf

IAEA (2007). *Lessons Learned from the Decommissioning of Nuclear Facilities and the Safe Termination of Nuclear Activities. Proceedings of an International Conference held in Athens, 11-15 December 2006*. IAEA Proceedings Series. http://www-pub.iaea.org/MTCD/Publications/PDF/Pub1299_web.pdf

IAEA (2008). *Long Term Preservation of Information for Decommissioning Projects*. Technical Reports Series No. 467. http://www-pub.iaea.org/MTCD/publications/PDF/trs467_web.pdf

IAEA (2009a). *An overview of stakeholder involvement in decommissioning*. IAEA Nuclear Energy Series No. NW-T-2.5. http://www-pub.iaea.org/MTCD/Publications/PDF/Pub1341_web.pdf

IAEA (2009b). *IAEA Safety Standards. Classification of Radioactive Waste*. General Safety Guide No.GSC 1. http://www-pub.iaea.org/MTCD/publications/PDF/Pub1419_web.pdf

IAEA (2010). *Nuclear Power Reactors in the World*. Reference Data Series No. 2. http://www-pub.iaea.org/MTCD/publications/PDF/iaea-rds-2-30_web.pdf

IAEA (2011a). *Redevelopment and Reuse of Nuclear Facilities and Sites: Case Histories and Lessons Learned*, IAEA Nuclear Energy Series No. NW-T-2.2.

IAEA (2011b). *Design Lessons Drawn from the Decommissioning of Nuclear Facilities*. IAEA TECDOC 1657. http://www-pub.iaea.org/MTCD/Publications/PDF/TE_1657_web.pdf

IAEA (2011c). International Conventions and Agreements. http://www.iaea.org/Publications/Documents/Conventions/jointconv.html

IAEA (2011d). The The Net-Enabled Radioactive Waste Management Database (NEWMDB). http://newmdb.iaea.org/datacentre-comparee.aspx

IAEA (2012a). Power Reactor Information System web site. http://www.iaea.org/programmes/a2/ (date retrieved: 20 January 2012)

IAEA (2012b). International Decommissioning Network (IDN). Waste Technology Section. http://www.iaea.org/OurWork/ST/NE/NEFW/WTS-Networks/IDN/overview.html

ICRP (International Commission on Radiological Protection) (2007). *The 2007 Recommendations of the International Commission on Radiological Protection*. ICRP Publication 103. http://www.icrp.org/docs/ICRP_Publication_103-Annals_of_the_ICRP_37(2-4)-Free_extract.pdf

Kazimi, M.S. (2003). Thorium Fuel for Nuclear Energy. *American Scientist*, 91, 408 http://www.americanscientist.org/issues/feature/2003/5/thorium-fuel-for-nuclear-energy/1

Kværner Moss Technology as (1996). Disposal of Russian nuclear submarines, 19 January. http://spb.org.ru/bellona/ehome/russia/nfl/nfl6.htm

McCombie, C. (2010). Spent fuel challenges facing small and new nuclear programmes, IAEA Conference on Management of Spent Fuel, June 2010

Mount, M.E., Sheaffer, M.K. and Abbott, D.T. (1994). Kara Sea radionuclide inventory from naval reactor disposal. *Journal of Environmental Radioactivity*, 25, 11-19

Nilsen, T., Kudri, I. and Nikitin, A. (1997). The Russian Northern Fleet: Decommissioning of Nuclear Submarines. Bellona Report No. 296

OECD/NEA (Organisation for Economic Co-operation and Development/Nuclear Energy Agency) (2010a). *Applying Decommissioning Experience to the Design and Operation of New Nuclear Power Plants*. NEA No. 6924. http://www.oecd-nea.org/rwm/reports/2010/nea6924-applying-decommissioning.pdf

OECD/NEA (2010b). *Public Attitudes to Nuclear Power*. NEA No. 6859. http://www.oecd-nea.org/ndd/reports/2010/nea6859-public-attitudes.pdf

OECD/NEA (2010c). Cost Estimation in Decommissioning: *An International Overview of Cost Elements, Estimation Practices and Reporting Requirements*. NEA No. 6831. http://www.oecd-nea.org/rwm/reports/2010/nea6831-cost-estimation-decommissioning.pdf

OECD/NEA (2010d). *Towards Greater Harmonization of Decommissioning Cost Estimates*. NEA No. 6867. http://www.oecd-nea.org/rwm/reports/2010/nea6867-harmonisation.pdf

OECD/NEA (2011). Forum on Stakeholder Confidence web site. http://www.oecd-nea.org/rwm/fsc/

Okyar, H.B. (2011). International survey of government decisions and recommendations following Fukushima. NEA News No. 29.2. http://www.oecd-nea.org/rwm/reports/2010/nea6867-harmonisation.pdf

O'Sullivan, O., Nokhamzon, J.G. and Cantrel, E. (2010). Decontamination and dismantling of radioactive concrete structures. NEA News No. 28.2. http://www.oecd-nea.org/nea-news/2010/28-2/NEA-News-28-2-8-updates.pdf

SEPA (Scottish Environment Protection Agency) (2011). Remediation of Radioactively Contaminated Sites. http://www.sepa.org.uk/about_us/sepa_board/agendas_and_papers/20_sept_2011.aspx

Shimada, T., Oshima, S. and Sukegawa, T. (2010). Development of Safety Assessment Code for Decommissioning of Nuclear Facilities (DecDose). *Journal of Power and Energy Systems,* 4(1), 40-53

SwissInfo (2011). Decommissioning nuclear plants comes at a price, 6 April. http://www.swissinfo.ch/eng/politics/Decommissioning_nuclear_plants_comes_at_a_price.html?cid=29936460

Szilagyi, A. (2012). Andrew Szilagyi, Director, Office of Decommissioning and Facility Engineering, United States Department of Energy, personal communication (26 January).

Tetra Tech (2012). Rocky Flats Decontamination and Demolition Project, CO. Tetra Tech EC, Inc. http://www.tteci.com/tteci/Department-of-Energy/rocky-flats-decontamination-and-demolition-project-co.html

UK DTI (Department of Trade and Industry, United Kingdom) (2002). Managing the Nuclear Legacy – A strategy for action

UK NDA (Nuclear Decommissioning Authority, United Kingdom) (2011). The Nuclear Legacy web site http://www.nda.gov.uk/aboutus/the-nuclear-legacy.cfm

US DOE (United States Department of Energy) (2009). *US DOE EM Strategy and Experience for In Situ Decommissioning*. Prepared By U.S. Department of Energy, Office of Environmental Management. Office of Engineering and Technology, EM-20. http://www.em.doe.gov/EM20Pages/PDFs/ISD_Strategy_Sept_4_2009.pdf

US DOE (2010). Awards Contract for Decontamination and Decommissioning Project for the East Tennessee Technology Park. Media release, 29 April. http://www.em.doe.gov/pdfs/DOE%20Awards%20Contract%20for%20ETTP%20DD.pdf

US DOE (2012). Office Of Environmental Management (EM): Deactivation & Decommissioning (D&D) Program Map. http://www.em.doe.gov/EM20Pages/DDMaps.aspx

US EPA (United States Environmental Protection Agency) (2011a). Hanford – Washington. http://yosemite.epa.gov/R10/CLEANUP.NSF/sites/Hanford

US EPA (2011b). Radiation Protection. http://epa.gov/rpdweb00/index.html

US GAO (United States Government Accountability Office) (2010). *Department of Energy: Actions Needed to Develop High-Quality Cost Estimates for Construction and Environmental Cleanup Projects*. Report to the Subcommittee on Energy and Water Development, Committee on Appropriations, House of Representatives. GAO-10-199. http://www.gao.gov/new.items/d10199.pdf

US NRC (United States Nuclear Regulatory Commission) (2001). NRC Seeks Early Public Comment on a Proposal to Permit Entombment for Reactor Decommissioning. NRC News No. 01-121, 11 October. http://pbadupws.nrc.gov/docs/ML0201/ML020150438.pdf

US NRC (2009). *Backgrounder on the Three Mile Island Accident* (page last reviewed/updated 15 March 2011). http://www.nrc.gov/reading-rm/doc-collections/fact-sheets/3mile-isle.html

US NRC (2011). Low-Level Waste Disposal. http://www.nrc.gov/waste/llw-disposal.html

WNA (World Nuclear Association) (2011a). Decommissioning Nuclear Facilities. http://www.world-nuclear.org/info/inf19.html

WNA (2011b). Nuclear power plants under construction. http://world-nuclear.org/NuclearDatabase/rdresults.aspx?id=27569&ExampleId=62

WNA (2011c). Nuclear Power in Gemany. http://world-nuclear.org/info/inf43.html

WNA (2011d). Nuclear-Powered Ships. http://www.world-nuclear.org/info/inf34.html

WNA (2011e). Waste Management. http://www.world-nuclear.org/education/wast.htm

WNA (2011f). Safe Decommissioning of Civil Nuclear Industry Sites. http://world-nuclear.org/reference/position_statements/decommissioning.html

WNN (World Nuclear News) (2011a). Decommissioning campaign complete at UK reactor, 16 June. http://www.world-nuclear-news.org/WR-Decommissioning_campaign_complete_at_UK_reactor-1606117.html

WNN (2011b). Fukushima units enter decommissioning phase, 21 December. http://www.world-nuclear-news.org/WR-Fukushima_units_enter_decommissioning_phase-2112114.html

WNN (2011c). Chubu agrees to Hamaoka shut down, 9 May. http://www.world-nuclear-news.org/RS-Chubu_agrees_to_Hamaoka_shut_down-0905115.html

WNN (2011d). US nuclear regulator Oks Vermont Yankee extension, 11 March. http://www.world-nuclear-news.org/newsarticle.aspx?id=29618

WNN (2012a). Japanese reactors await restart approvals, 16 January. http://www.world-nuclear-news.org/RS-Japanese_reactors_await_restart_approvals-1601124.html

WNN (2012b). Vermont Yankee wins right to keep generating, 20 January. http://www.world-nuclear-news.org/RS_Vermont_Yankee_wins_right_to_keep_generating_200112a.html

Wood, J. (2007). *Nuclear Power*. Institution of Engineering and Technology (IET). Power and Energy Series No. 52.

Wuppertal (ed.) (2007). *Comparison among Different Decommissioning Funds Methodologies for Nuclear Installations: Country Report France. On Behalf of the European Commission Directorate-General Energy and Transport*. Wuppertal Institut für Klima, Umwelt, Energie GmbH im Wissenschaftszentrum Nordrhein-Westfalen (Wuppertal Institute for Climate, Environment and Energy, Science Centre North Rhine-Westphalia
http://www.wupperinst.org/uploads/tx_wiprojekt/EUDecommFunds_FR.pdf

Yanukovych, Viktor, President of Ukraine: official website (2011). President visits Chornobyl Nuclear Power Plant. http://www.president.gov.ua/en/news/19883.html

Key Environmental Indicators
Tracking progress towards environmental sustainability

Measuring changes in the global environment and keeping track of these changes is an important first step in raising awareness and addressing issues of concern. Although, as Einstein said, "not everything that can be counted counts, and not everything that counts can be counted", today we often need to measure and monitor before we can identify and manage problems. Climate change would not have been recognized as a major issue if we had not had solid time-series data on air temperatures and melting glaciers. Many people argue that ecosystems and biodiversity are not being managed properly because they are undervalued, and therefore are not adequately reflected in economic systems and accounting mechanisms.

Detailed measurement data can be translated into more readily understandable indicators and presented in clear graphics. These indicators and graphics help to explain the phenomena we see around us, and ultimately help to define policies and actions with which to respond to unfavourable trends. A good map or graph is "worth more than a thousand words". At the same time, indicators are not more than that – they illustrate trends regarding phenomena measured over time.

This chapter presents major global environment trends using a small number of key indicators. It draws attention to major issues of concern with respect to air, water, land and biodiversity and so helps us understand where the environment stands. Keeping track of such trends on a yearly basis is of fundamental importance in order to make sure the world is well-informed, raise awareness, and support national and international decision-making processes. More in-depth studies are often needed to gain better insight into the dynamics and complexity of environmental issues and underlying causes, so as to enable the development of effective management strategies and concrete policy actions.

Where possible, the indicators presented in this chapter coincide with those identified in the GEO assessment process. The fifth Global Environment Outlook (GEO-5) is a comprehensive, integrated assessment of the world's environment. It not only provides a review of the state of the environment, but it analyses policies that work and puts forward options and pathways for reaching a more sustainable world. Also highlighted are indicators that are part of the set of indicators to track progress towards reaching the Millennium Development Goals (MDGs) as defined under the UN Millennium Declaration, a global initiative to foster sustainable development.

The set of key indicators in this chapter constitute a snapshot of major global and regional environmental issues, to the extent that data are available. In several cases, data are insufficient or too incomplete to properly illustrate at aggregate levels what the numerical trends are. Notorious examples include trends related to the use of chemicals, waste collection, freshwater quality, urban air pollution, biodiversity loss and land degradation.

Depletion of the ozone layer

The Montreal Protocol on Substances that Deplete the Ozone Layer has served as an effective instrument for protecting the stratospheric ozone layer. It provides an international framework

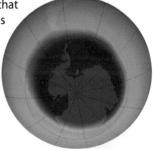

The ozone hole over the Antarctic in September 2011. The largest Antarctic ozone hole was measured in September 2006. Full recovery of the Antarctic ozone layer is not expected until after 2050. *Credit: NOAA*

◄ A remote environmental monitoring unit from the Woods Hole Oceanographic Institution is lowered in Panama waters. *Credit: John F. Williams, US Navy*

Indicators are measures that can be used to illustrate and communicate complex phenomena in a simple way, including trends and progress over time.

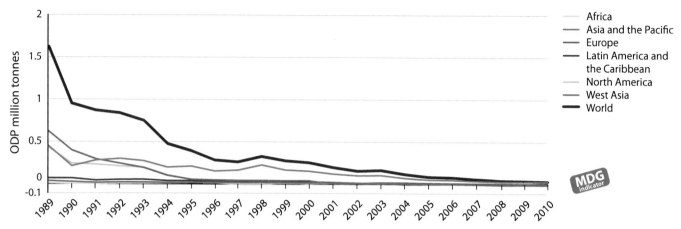

Figure 1: Consumption of ozone-depleting substances expressed as million tonnes of ozone depletion potential (ODP), 1989-2010. Although challenges remain, consumption of ozone-depleting substances has declined tremendously thanks to the Montreal Protocol to the Vienna Convention for the Protection of the Ozone Layer. Source: GEO Data Portal, compiled from UNEP (2011a)

for phasing out ozone-depleting substances (ODS), including chlorofluorocarbons (CFCs) and hydrochlorofluorocarbons (HCFCs). About 98 per cent of all ozone-depleting substances controlled under the Protocol have been phased out (**Figure 1**). As a result, the ozone layer is expected to return to its pre-1980 levels around the middle of this century. However, products being used as substitutes may also have significant effects on climate change. Hydrofluorocarbons (HFCs) are excellent alternatives for use in refrigerators and industrial air conditioners. But while they do not deplete stratospheric ozone, they are extremely powerful greenhouse gases with a high global warming potential. Globally, HFC emissions are currently growing at a rate of 8 per cent per year (**Figure 2**). The use of HFCs could potentially wipe out all the climate benefits gained through phasing out CFCs and other ozone-depleting substances (UNEP 2011b).

Early in 2011, unprecedented ozone losses were reported over the Arctic (Manney et al. 2011). Scientists attributed this phenomenon to unusual long-lasting cold conditions that contributed to the processes depleting the stratospheric ozone layer.

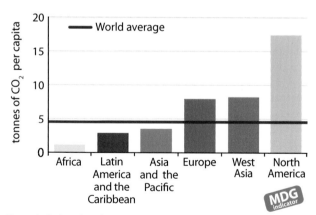

Figure 2: Consumption of HFCs in gigagrams, 1968-2008. Substitutes for ozone-depleting substances, such as HFCs, can have significant impacts on climate change. Source: JRC/PBL (2010), UNEP (2011b)

Figure 3: Carbon dioxide emissions per capita, 2008. Per capita emissions of CO_2 are well above the global average in Europe, West Asia and, most notably, North America. Source: GEO Data Portal, compiled from Boden et al. (2011)

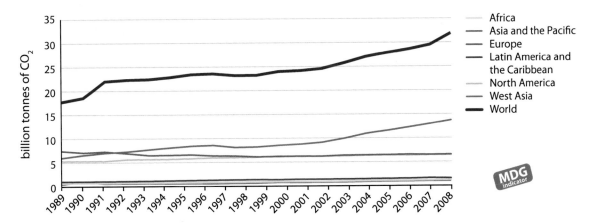

Figure 4: Carbon dioxide emissions from fossil fuels and cement production, expressed in billions of tonnes of CO_2, 1989-2008. Global CO_2 emissions have increased in recent years, mainly in the Asia and the Pacific region. *Source: GEO Data Portal, compiled from Boden et al. (2011)*

Climate change

Carbon dioxide (CO_2) emissions from the burning of fossil fuels are a major contributor to climate change. Per capita emissions of CO_2 continue to be highest in North America, followed by West Asia and Europe, and are lowest in Africa **(Figure 3)**. Global CO_2 emissions are continuing to increase, reaching 32.1 billion metric tonnes in 2008, an increase of 2.4 per cent compared with the previous year and 42 per cent compared with 1990 **(Figure 4)**. The level of CO_2 emissions differs greatly among regions and countries. In the last decade the increase has been most significant in the Asia and the Pacific region. With increasing emissions, CO_2 concentrations in the atmosphere have gone from an estimated 280 ppm in pre-industrial times, and 315 ppm in 1958, to 390 ppm in 2011, causing global warming (Tans and Keeling 2011). Despite short-term spatial and temporal variability, a long-term trend of global warming can be seen. The past decade was the warmest on record since 1880 in terms of average global temperatures. The ten warmest years on record have all occurred since 1998 (UNEP 2011c).

One of the clearest signals of global warming is the melting of glaciers in several parts of the world **(Figure 5)**. The rapid, possibly accelerated, melting and retreat of glaciers has severe impacts on water and energy supply, sea level fluctuations, vegetation patterns, economic livelihoods and the occurrence of natural disasters. Dramatic glacier shrinkage may lead to the deglaciation of large areas of many mountain ranges by the end of this century (WGMS 2008).

Fossil fuels such as oil, coal and gas continue to dominate global energy supply **(Figure 6)**. Notwithstanding gains in energy efficiency and greater use of renewable energy sources, total use of fossil fuels currently makes up about 80 per cent of the primary energy supply. However, global investment in renewable energy is growing sharply. It stood at US$211 billion (thousand million) in 2010, more than five times the amount in 2004 **(Figure 7)**. While the overall share of renewable energy is currently just over 13 per cent, there has been a spectacular increase in the use of solar and wind energy, as well as of biofuels, in recent years **(Figure 8)**.

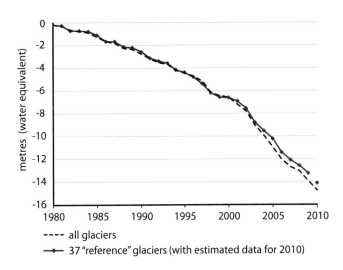

Figure 5: Mountain glacier mass balance. Glaciers continue to melt at unprecedented rates, producing increasingly severe impacts on the environment, natural resources and human well-being. *Source: WGMS (2011)*

Carbon trading is a relatively new instrument whose use has grown rapidly (**Figure 9**). Following five consecutive years of robust growth, the carbon market reached a three-year plateau between 2008 and 2010 with a value of around US$140 billion. This equals about 0.2 per cent of global GDP. The increase with respect to global market trends since 2005 – the year the Kyoto Protocol entered into force – is mostly due to an increase in transactions volume. Carbon prices have not been unaffected by the recent economic downturn. In a period of less than a year, prices fell from €30 to €8 on the European market. Moreover, due to a lack of clarity about regulations in the post-Kyoto regime after 2012, some of the implemented mechanisms suffer today from large losses in value. Out of the total amount of allowances, the EU Emissions Trading System launched in 2005 accounted for 84 to 97 per cent of the global carbon market value in 2010 (World Bank 2011).

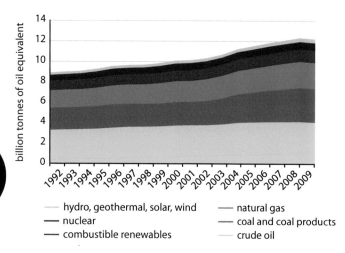

Figure 6: Primary energy supply, 2009. Use of fossil fuels has increased steadily over the past two decades, although there has been a levelling off in recent years. Renewable resources represent a modest but rising share. *Source: IEA (2011a), REN21 (2011)*

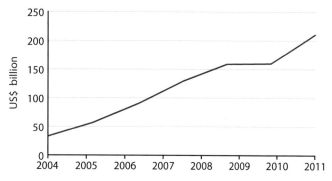

Figure 7: Investment in renewable energy, 2004-2011. Global investment in renewable energy has grown rapidly in recent years. It stood at US$211 billion in 2011. *Source: UNEP (2011d)*

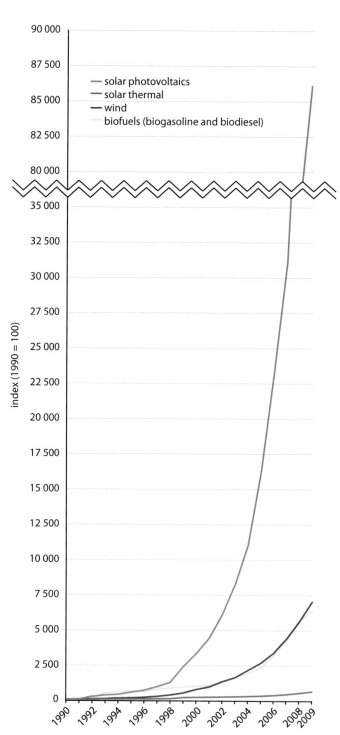

Figure 8: Renewable energy supply index, 1990-2009 (1990=100). Use of solar energy is on the rise – even skyrocketing – followed by wind and biofuels. *Source: GEO Data Portal, compiled from IEA (2011b)*

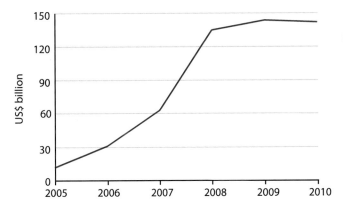

Figure 9: Carbon market growth in US$ billion (thousand million). The carbon market has reached around US$140 billion in recent years, mainly due to an increase in transactions, while prices have fallen as a result of the economic downturn. *Source: World Bank (2011)*

The level of small particles in the air (PM_{10}), which affect both global warming and human health, still far exceeds the World Health Organization's recommended maximum level of 20 µg/m³ in several large cities (WHO 2006, 2011). They include Beijing, Cairo and New Delhi. The air pollution map shows a high level of very small particles ($PM_{2.5}$), notably in parts of Asia, West Asia and Africa (**Figure 10**). Nevertheless, data on small particles should be treated with care, as these are estimates based on models and are sensitive to local conditions.

Natural resource use

The depletion of natural resources continues in many ways and in many parts of the world. Water, land and biodiversity are under great pressure almost everywhere. Exploitation of fish stocks is an example (FAO 2011a). It is estimated that the percentage of overexploited, depleted or recovering stocks has been increasing for many years, reaching 33 per cent in 2008 – to the detriment of underexploited or moderately exploited stocks (**Figure 11**). The marine fish catch has leveled off in recent years except in the Asia and the Pacific region, where it continues to rise. Aquaculture has increased significantly, again mostly in Asia and particularly in China (**Figure 12**). By 2009, global aquaculture production had risen to 51 million tonnes while the global total fish catch remained below 90 million tonnes. Aquaculture has significant benefits for many people and economies, but there are disadvantages: among other impacts, large quantities of wild-caught fish are used for feed, mangroves in coastal areas are lost when fish farms are created, and significant amounts of chemicals and pharmaceuticals (including antibiotics) may be used and discharged to the environment (FAO 2011b).

Pressures on marine and coastal ecosystems are further increased by progressive ocean acidification resulting from higher levels of CO_2 in the atmosphere (**Figure 13**). As atmospheric CO_2 increases, the oceans absorb more of it,

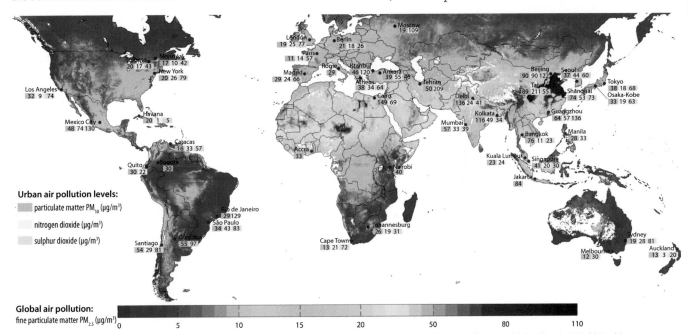

Figure 10: Global and urban air pollution levels. Estimated data indicate that air pollution is well above the guidelines established by the World Health Organization in many large cities in the world, for instance in terms of small particles, particularly in developing regions. *Sources: van Donkelaar et al. (2010), World Bank (2010)*

KEY ENVIRONMENTAL INDICATORS

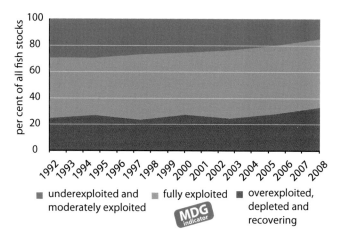

Figure 11: Fish stocks exploitation. The percentage of fish stocks fully exploited, overexploited, depleted or recovering has increased to 85 per cent. Source: GEO Data Portal, compiled from FAO (2011a)

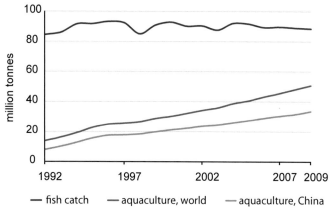

Figure 12: Fish catch and aquaculture production. While the global fish catch has stabilized at around 90 million tonnes per year, aquaculture has been increasing significantly, particularly in China and other parts of Asia. Source: FAO (2011b)

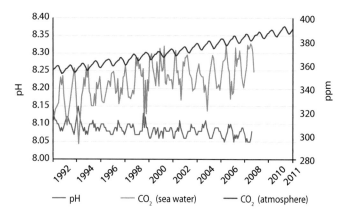

Figure 13: Atmospheric CO_2 concentrations and ocean acidification, indicated by increased partial pressure of CO_2 and lower pH of global mean surface water. Source: Caldeira and Wickett (2003), Feely et al. (2009), Tans and Keeling (2011)

increasing the partial pressure of CO_2 and causing a decrease in pH. An increase in ocean acidification can have significant consequences on marine organisms, which may alter species composition, disrupt marine food webs and ecosystems, and potentially damage fishing, tourism and other human activities connected with the sea. Of particular concern are corals, shell fish and skeleton-forming phytoplankton. Given these trends, the clock is ticking on the sustainability of global fish stocks and marine biodiversity, and the need for an international agreement on better management of the marine environment is growing more urgent (UNEP 2010).

Although the overall rate of deforestation is slowing down, large forest areas are still declining, particularly in Latin America and Africa (**Figure 14**). At the same time, the total area under forest plantations has been increasing steadily, with much of this area devoted to the cultivation of oil palm for food production and of biofuel crops (**Figure 15**).

The total forest area managed under the two largest forest certification bodies – the Forest Stewardship Council (FSC) and the Programme for Endorsement of Forest Certification (PEFC) – has increased by an impressive 20 per cent per year since 2002 (**Figure 16**). However, the total area under any of these schemes is still modest and currently represents about 10 per cent of all forests, mainly in Europe and North America. Similarly, the extent of protected areas has been has been increasing gradually in all regions of the world (**Figure 17**). However, the extent of marine protected areas remains low, with only 7 per cent of coastal waters and 1.4 per cent of oceans protected. New global targets have been set for the extent of protected areas. Governments agreed in 2010 to protect 17 per cent of terrestrial and inland waters, and 10 per cent of coastal and marine areas, by 2020 (CBD 2010).

Biodiversity loss continues to be an issue of major concern, as indicated by the Red List Index (RLI) of Threatened Species (**Figure 18**). The Red List measures the risk of extinction of species in seven classes, ranging from Least Concern to Extinct. A value of 1.0 indicates that species are not expected to become extinct in the near future, while 0.0 means that a species is extinct. A small change in the level of threat can have significant impacts on species decline. In the case of species groups and years for which data are available (i.e. birds, mammals and amphibians since 1992), the trend is downward. Almost one-fifth of vertebrate species are classified as threatened, ranging from an estimated 13 per cent of birds to 41 per cent of amphibians. The highest number of threatened vertebrates are found in the tropical regions (Hoffman et al. 2010).

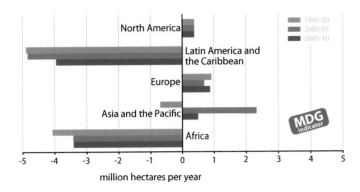

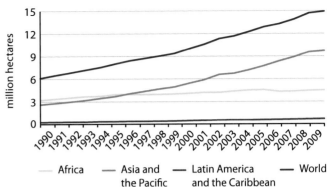

Figure 14: Forest cover annual change. The overall deforestation rate is decreasing and, in some regions, forested areas are increasing. Still, large areas of forest are declining, particularly in Latin America and Africa. *Source GEO Data Portal, compiled from FAO (2010)*

Figure 15: Area of oil palm harvested. Increases in global demand for food and fuel are driving forest clearance in the tropics. Much of this forest loss is due to rapid expansion of oil palm monocropping. *Source: GEO Data Portal, compiled from FAO (2011c)*

The regulation and reporting of international trade in endangered species is increasing significantly, as recorded by the Convention on International Trade in Endangered Species of Wild Fauna and Flora (CITES). This is partially due to the growing number of parties to the convention and to the number of species included in the Appendices. CITES aims to ensure that international trade in animal and plant species listed under the Convention is legal, sustainable and traceable. Types of trade are diverse, ranging from live animals and plants to a wide range of wildlife products derived from them, including food products, exotic leather goods, wooden musical instruments, timber, tourist curios and medicines. Each year, international wildlife trade is estimated to be worth billions of dollars and to include hundreds of millions of plant and animal specimens. The trade volume reveals that reported trade in live animals increased up to the first half of the 1990s, thereafter remaining reasonably

Figure 16: Forest certification by the Forest Stewardship Council (FSC) and the Programme for Endorsement of Forest Certification schemes (PEFC) in 2011. There has been an impressive increase in forest certification, but it is largely taking place in Europe and North America. *Source: FSC (2012), PEFC (2012)*

KEY ENVIRONMENTAL INDICATORS

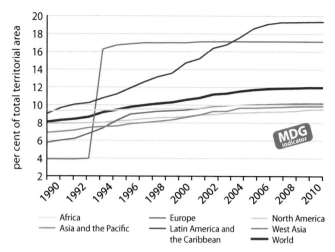

Figure 17: Ratio of area protected to maintain biological diversity to surface areas. The extent of protected areas has been increasing, particularly in Latin America and the Caribbean but also in West Asia after a large single protected area was created in 1994. Source: GEO Data Portal, compiled from UNEP-WCMC (2011)

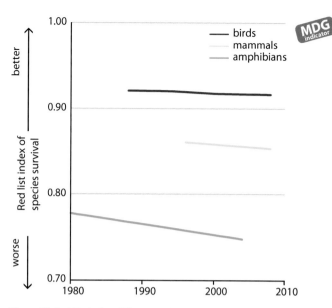

Figure 18: Red List Index of Threatened Species. Species groups such as birds, mammals and amphibians are increasingly threatened. Although data are insufficient, the status of other groups is likely to be similar if not worse. Source: IUCN (2011)

stable until 2005 and then declining in recent years (**Figure 19**). However, the proportion of captive bred specimens has increased and, in recent years, these specimens have generally outnumbered wild animals reported in trade.

Significant progress has been made in improving access to clean drinking water, with the global figure approaching 90 per cent in 2010 (**Figure 20**). In some regions, such as Africa and Asia and the Pacific, the increase has been remarkable although challenges remain, especially in rural areas. Nevertheless, the world is far from reaching the target for access to improved sanitation (WHO/UNICEF 2012). Even as progress is reported in all parts of the world, about half the population of developing regions does not use improved sanitation. In this case, too, urban areas are better serviced than rural ones although disparities are decreasing (WHO/UNICEF 2012).

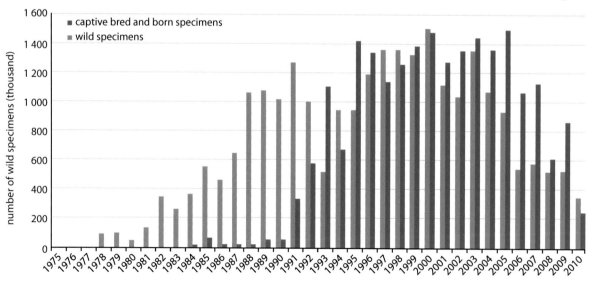

Figure 19: Trade in captive bred and born specimens versus wild specimens, 1975-2010. Regulation and reporting of trade in live animals has increased considerably, with trade in captive bred animals exceeding trade in wild animals in recent years. Source: CITES (2012)

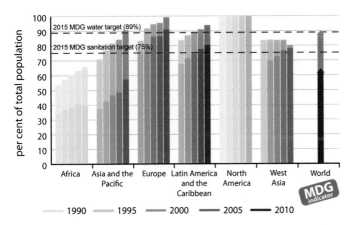

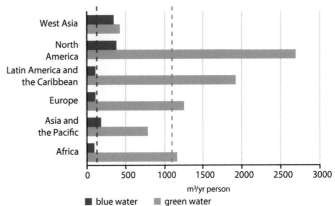

Figure 20: Proportion of the population with sustainable access to an improved water source and with access to improved sanitation. Preliminary data are included for 2010. The global MDG target for drinking water will be exceeded by 2015, but the target for sanitation will be missed. Challenges remain in many parts of the world, particularly in rural areas in developing regions. *Source: GEO Data Portal, compiled from WHO/UNICEF (2012)*

Figure 22: Water footprint, blue and green. A country's water footprint is the total volume of freshwater used to produce the goods and services consumed by its population. The footprint for "blue water", related to consumption of surface and groundwater resources, is highest in North America and Latin America and the Caribbean. The footprint for "green water", related to use of rainwater, is highest in North America and West Asia. *Source: Mekonnen and Hoekstra (2011)*

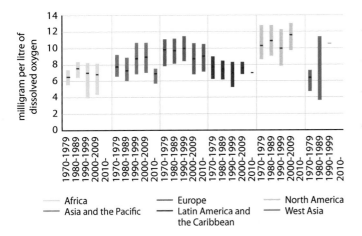

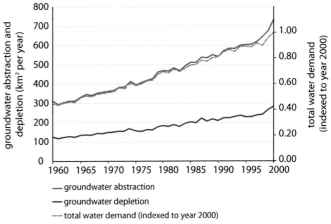

Figure 21: Levels of dissolved oxygen in surface waters. Available data appear to indicate that concentrations of dissolved oxygen are generally within the widely accepted limits of between 6 mg/l in warm water and 9.5 mg/l in cold water, as set, for instance, in Australia, Brazil and Canada. However, it should be noted that these data are not representative of all waters in the regions, or of each decade, depicted here. *Source: UNEP-GEMS/Water (2011)*

Figure 23: Groundwater abstraction and depletion in km³ per year, 1960-2000. Groundwater depletion has increased steadily in the past decades, along with demand and withdrawals, as indicated by modelled data. *Source: Wada et al. (2010)*

Uncontrolled discharges of sewage to surface water has a direct impact on water quality. Levels of dissolved oxygen in surface waters are a good indicator of environmental conditions for aquatic life. Eutrophication or nutrient over-enrichment may raise the concentrations due to increased productivity from phytoplankton, while organic pollutants would increase oxygen demand and lower the concentration (**Figure 21**). Whereas the effects are often local, the cumulative impact on the quality of freshwater bodies is being acknowledged as a major global concern (UNEP 2012a). Water quality monitoring is well-established in some regions, but is far from adequate in others. Few possibilities to provide a global picture of water quality exist, largely due to data gaps, data access, and limited capabilities and resources.

A country's total water footprint is the total volume of freshwater used to produce the goods and services consumed by that country's population. It may partially originate outside the country (**Figure 22**). Apart from water use for human

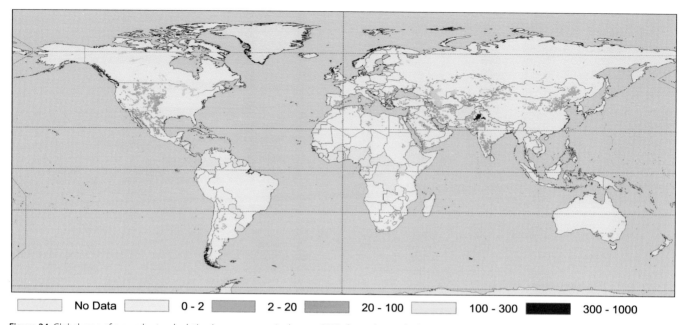

Figure 24: Global map of groundwater depletion in mm per year in the year 2000. Groundwater depletion remains most severe in parts of Asia and the Pacific, as well as in West Asia and North America. *Source: Wada et al. (2010)*

consumption ("blue water"), water is needed to sustain ecosystems and the services they provide to society ("green water"). The total water footprint has been increasing in many regions, with many countries significantly externalizing their water footprint by importing water-intensive goods, thus putting pressure on water resources in the exporting regions (Mekonnen and Hoekstra 2011). The main use of water is in the agricultural sector, followed by industry and households.

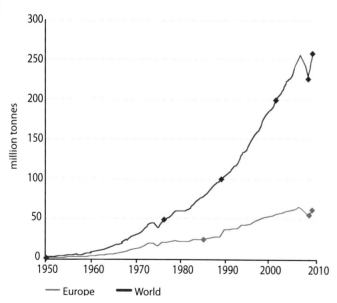

Figure 25: Plastics production in million tonnes, 1950-2010. After a dip around 2008-2009, world production reached a new record of 265 million tonnes in 2010. Plastic debris in the ocean has become an issue of growing concern in recent years. *Source: PlasticsEurope (2011)*

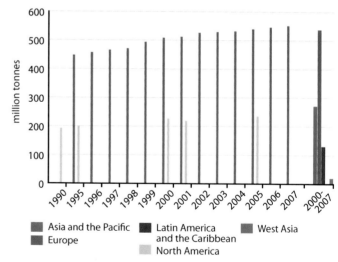

Figure 26: Municipal waste collection in different regions, 1990-2007. Insufficient data are available for Africa. The collection of waste is highest in Europe, followed by Asia and the Pacific and North America. Solid data on the generation, collection and treatment of hazardous and other wastes are still very sparse in most parts of the world. *Source: GEO Data Portal, compiled from UNSD (2011), EPA (2012)*

In Asia's largest slum, Dharavi, India, almost 80 per cent of dry waste, such as plastic, paper and scrap, is separated for recycling. *Credit: Cristen Rene*

The growth of the human water footprint is also reflected in the amounts of groundwater extracted. Groundwater use increased steadily between 1960 and 2000 (**Figure 23**). The 2000 map of groundwater depletion illustrates areas where groundwater abstraction exceeds replenishment, causing groundwater depletion (**Figure 24**). The resulting lowering groundwater tables and insufficient groundwater fluxes may put groundwater dependent ecosystems at risk of dessication and cause harm to regulating functions and other ecosystem services.

Chemicals and waste

The amount and number of chemicals and waste that end up in our environment are increasing. As discussed in the previous edition of the *UNEP Year Book*, plastic debris ending up in the ocean is of growing concern because of its possible chemical impacts (UNEP 2011d). The production of plastics is a proxy for the amount of plastic debris that may eventually find its way to waterways and the ocean (**Figure 25**). The solution to this and many other waste problems lies in better waste management. Unfortunately, reliable and comparable data on the generation, collection and management of waste are very scarce and vary widely across and within regions. Some progress can be noted with respect to data on the generation of hazardous and other wastes, but definite trends for the various regions and the world as a whole cannot be presented. The amount of waste collected by municipalities (**Figure 26**) allows an indicative comparison among regions except for Africa. Municipal waste collection is highest in Europe, with the amount having increased steadily to about 552 million tonnes in 2007. Based on average figures for the 2002-2009 period, in other regions the amount of waste collected is less. For Africa there are no regional waste data available.

Environmental governance

International environmental governance (IEG) has come to the forefront in the debate on how to achieve sustainable development. In the context of the institutional framework for sustainable development, it is one of the issues to be discussed at the 2012 UNCSD in Rio de Janeiro (Rio+20). Among the major global instruments are Multilateral Environmental Agreements (MEAs) covering climate, biodiversity, chemicals and other issues. There have been significant increases in the number of countries that have become a Party to these agreements and conventions (**Figure 27**).

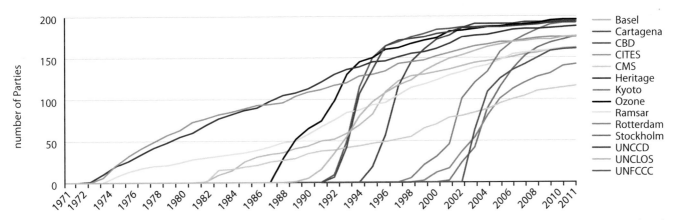

Figure 27: Number of Parties to Multilateral Environmental Agreements (MEAs), 1971-2011. Many MEAs and conventions are reaching the maximum number of countries as signatories (Parties). Taking all 14 MEAs depicted here together, the number of Parties reached 89 per cent in 2011. Establishing and signing such agreements is a first important step, but it does not mean the environmental problems addressed will be solved right away. *Sources: Compiled from various MEA secretariats (see reference list)*

Signing and ratifying an international agreement or protocol does not mean that appropriate measures are being put in place or that an environmental problem is on its way to being solved. It does demonstrate that there is awareness and a commitment to address prominent issues. Several agreements have been very successful, such as the Montreal Protocol on the Depletion of the Ozone Layer. The majority of the countries in the world are now a Party to many of the global MEAs, and some MEAs have approached the maximum number of Parties at close to 200. Taken together, the 14 MEAs shown in the figure saw the number of Parties increase to 89 per cent in 2011, up from 69 per cent since the last MEA (the Cartagena Protocol on Biosafety (Cartegana Protocol on Biosafety) entered into force in 2003.

From a more business-oriented viewpoint, it is possible to look at the trend in certifications for environmental management such as ISO 14001 (**Figure 28**). ISO 14001 codifies the practices and standards that companies and other organizations should follow in order to minimize the harmful effects of their activities on the environment and improve their environmental performance. Certification indicates the extent of conformity with environmental policies as stated by the companies. It does not necessarily mean that performance is improved or environmental impacts are reduced. The strong increase in the number of ISO 14001 certifications, with more than 18 times as many certifications in 2010 as in 1999, shows a growing commitment by companies and organizations to adopt environmental management systems.

Looking ahead

In the past 20 years the world has experienced changes in economic production and consumption patterns, international trade, and information and communication technologies. Significant changes have also occurred in the environmental domain, with accumulating evidence of climate change and its impacts on the planet, of rapid biodiversity loss and species extinctions, of further degradation of land and soils and of the deterioration of inland waters and oceans. Environmental and other indicators enable us to keep track of the state of the environment. They will be used to inform the upcoming UNCSD in Rio de Janeiro (Rio+20) about progress since the original "Rio" summit, the UN Conference on Environment and Development (UNCED) in 1992 (UNEP 2011c).

Some progress has been made since 1992, including a significant reduction in the use of ozone-depleting substances, an increase in the use of renewable energy (notably solar and wind power), and the introduction of new mechanisms such as carbon trading and product certification. However, the overall picture presented by this set of key environmental indicators is not very positive. In the areas of climate change, biodiversity, glacier melt and fisheries, for instance, huge challenges remain with respect to addressing underlying causes and reversing trends.

Strengthening environmental governance is a cross-cutting issue. Strong environmental governance is critical to the achievement of environmental progress and sustainability. It is needed at all levels to respond quickly and effectively to emerging environmental challenges, and to work towards agreed environmental priorities. Some positive signs can be detected in this regard, including those related to addressing ozone depletion, the creation of protected areas, and the establishment of market mechanisms and certification schemes that put environmental issues at the core of economic thinking and decision making.

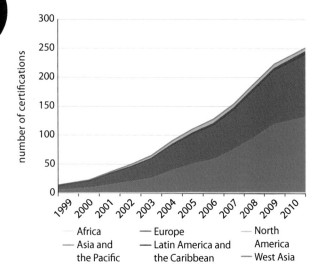

Figure 28: Number of ISO 14001 environmental management certifications, 1998-2010. ISO 14001 certification indicates that companies and other organizations are committed to adopt environmental management systems, in terms of confirming to their own stated policies. The total number of certifications surpassed 250 000 in 2010, with the highest shares in Asia and the Pacific and Europe. *Source: ISO (2011)*

The key environmental indicators in this chapter merely provide a snapshot of the global environment. The quality of some of these indicators is severely hampered by data gaps and other shortcomings in measurements. Providing solid time-series of major environmental trends at global and regional level is only possible in the case of a very limited number of indicators. Several trends cannot be properly presented here, or cannot be

Progress has been made since 1992, including the increasing use of renewables and new mechanisms such as carbon trading and product certification. Credit: FSC - Germany

presented at all, due to poor data availability and quality including lack of regional data or of long-term monitoring results. For some other issues, there have not been recent updates so that no meaningful new information can be presented.

A number of international initiatives and networks have been set up to look at key environmental indicators and underlying data issues and to help improve the data situation. They include those following progress on reaching the Millennium Development Goals (MDGs), the Group on Earth Observation and its System of Systems, the various Global Observation Systems, the Intersecretariat Working Group on Environment Statistics, and many others. A very practical effort that aims at "bridging the environmental data gap" is being made by the UN Statistical Division, in co-operation with UNEP, in the form of a biennial questionnaire on environment statistics, with the aim of collecting data on environmental topics from countries complementary to what is already covered by other international agencies, and connecting to capacity-building for data collection in developing regions (**Box 1**).

Such data efforts are critical to keeping the state of the environment under review. The *UNEP Year Book* will continue to provide the latest available information on an annual basis to support decision making and early identification of trends. For easy reference, an overview with the latest data of key indicators is presented at the end of this chapter (**Table 1**).

Box 1: The UNSD/UNEP Questionnaire on Environment Statistics

The biennial Questionnaire on Environment Statistics collects country-level data focusing on water and waste topics. In order to avoid duplication, this data collection effort covers countries which do not already report to the Joint OECD/Eurostat Questionnaire and addresses environmental topics not covered by other UN or or other international agencies.

The recent 2010 questionnaire was sent to 172 countries and territories, of which 87 (51 per cent) responded. Five of these reported not having data available. The best response rates were in Eastern Europe (79 per cent), followed by the Americas (62 per cent) and Asia (57 per cent). The response rate in Africa was 44 per cent and there was no response from Oceania. Among the 82 countries submitting data, 66 were able to provide data for both the water and waste sections of the questionnaire, while 16 countries provided data for only one of the two sections.

Following a thorough validation process, selected water and waste statistics with relatively good quality and geographic coverage (complemented by data from OECD/Eurostat) are published by UNSD as part of the Environmental Indicators and the Country Snapshots webpages (UNSD 2012a, b). The complete data and footnotes are uploaded to the Country Files webpage (UNSD 2012c). UNEP includes the data, together with (sub-)regional aggregations, in the GEO Data Portal (UNEP 2012b). Selected water and waste statistics are also updated in the "One UN" data entry point (UN 2012).

The water and waste statistics are essential to provide a sound picture of global and regional trends in these important sectors. However, data collection related to water and waste remains a challenge, often due to lack of capacity in countries. The next UNSD/UNEP environment data collection will take place in the course of 2012.

References

Basel (2012). Convention on the Control of Transboundary Movements of Hazardous Wastes and Their Disposal. http://www.basel.int/Countries/StatusofRatifications/PartiesSignatories/tabid/1290/Default.aspx

Boden, T.A., Marland, G. and Andres, R.J. (2011). Global, Regional, and National Fossil-Fuel CO_2 Emissions. Carbon Dioxide Information Analysis Center, Oak Ridge National Laboratory, U.S. Department of Energy, Oak Ridge, Tennesse, U.S.A. http:// cdiac.ornl.gov/trends/emis/tre_coun.html

Caldeira, K. and Wickett, M.E. (2003). Anthropogenic Carbon and Ocean pH. *Nature*, 425, 365

Cartagena (2012). Cartagena Protocol on Biosafety to the Convention on Biological Diversity http://bch.cbd.int/protocol/

CBD (2010). A new era of living in harmony with nature is born at the Nagoya Biodiversity Summit. Secretariat of the Biological Diversity Convention, Montreal

CBD (2012). Convention on Biological Diversity. http://www.biodiv.org/world/parties.asp

CITES (2012). Convention on International Trade in Endangered Species of Wild Fauna and Flora. http://www.cites.org/eng/disc/parties/

CMS (2012). Convention on the Conservation of Migratory Species of Wild Animals. http://www.cms.int/about/part_lst.htm

FAO (2010). Global Forest Resources Assessment 2010 (FRA). Key findings. Food and Agriculture Organization, Rome.

FAO (2011a). World Review of Fisheries and Aquaculture. Food and Agriculture Organization, Rome

FAO (2011b). Fisheries and Aquaculture Department: Global Statistical Collections. Food and Agriculture Organization, Rome. http://www.fao.org/fishery/statistics/en

FAO (2011c). FAO Stat database. Food and Agriculture Organization, Rome. http://faostat.fao.org

Feely, R.A., Doney, S.C. and Cooley, S.R. (2009). Ocean Acidification: Present Conditions and Future Changes in a High-CO_2 World. *Oceanography*, 22(4), 36-47

FSC (2012). Global FSC certificates. Forest Stewardship Council. http://www.fsc.org/ facts-figures.html

Heritage (2012). Convention Concerning the Protection of the World Cultural and Natural Heritage (World Heritage). http://whc.unesco.org/en/statesparties/

Hoffmann, M., Hilton-Taylor C., Angulo A. and others (2010). The Impact of Conservation on the Status of the World's Vertebrates. *Science*, 330(6010), 1503-1509

IEA (2011a). Energy balances for OECD and non-OECD countries (2011 edition). International Energy Agency, Paris. http://data.iea.org/ieastore/product.asp?dept_id=101&pf_id=309

IEA (2011b). Renewable Information (2011 edition). International Energy Agency, Paris. http://data.iea.org/ieastore/product.asp?dept_id=101&pf_id=309

ISO (2011). The ISO Survey of Certifications 2009. International Organization for Standardization, Geneva. http://www.iso.org/iso/iso_catalogue/management_standards/certification/the_iso_survey.htm

IUCN (2011). The IUCN Red List of Threatened Species (version 2010.4). International Union for Conservation of Nature. http://www.iucnredlist.org/about/summary-statistics

JRC/PBL (2010). Emission Database for Global Atmospheric Research (EDGAR), release version 4.2. Joint Research Centre (JRC)/PBL Netherlands Environmental Assessment Agency. http://edgar.jrc.ec.europe.eu

Kyoto (2012). Kyoto Protocol to the UN Framework Convention on Climate Change. http://unfccc.int/essential_background/kyoto_protocol/status_of_ratification/items/2613.php

Manney, G.L., Santee, M.L., Rex, M., Livesey, N.J., Pitts, M.C., Veefkind, P., Nash, E.R., Wohltmann, I. Lehmann, R., Froidevaux, L., Poole, L.R., Schoeberl, M.R., Haffner, D.P., Davies, J., Dorokhov, V., Gernandt, H., Johnson, B., Kivi, R., Kyrö, E., Larsen, N., Levelt, P.F., Makshtas, A., McElroy, C.T., Nakajima, H., Parrondo, M.C. and others (2011). Unprecedented Arctic ozone loss in 2011. *Nature*, 478, 469-475

Mekonnen, M.M. and Hoekstra, A.Y. (2011). The green, blue and grey water footprint of crops and derived crop products. *Hydrology and Earth System Sciences*, 15, 1577-1600.

Ozone (2012). Vienna Convention for the Protection of the Ozone Layer and its Montreal Protocol on Substances that Deplete the Ozone Layer (Ozone). http://ozone.unep.org/new_site/en/vienna_convention.php

PEFC (2012). Programme for the Endorsement of Forest Certification: Certified foresrt area by country. http://www.pefc.org/resources/webinar/item/801

PlasticsEurope (2011). Plastics – the facts. An analysis of European plastics production, demand and recovery for 2010 (updated on 4 November 2011).

Ramsar (2012). Convention on Wetlands of International Importance Especially as Waterfowl Habitat. http://www.ramsar.org/cda/en/ramsar-about-parties-contracting-parties-in-20715/main/ramsar/1-36-123%5E20715_4000_0__

REN21 (2011). Renewables 2011 Global Status Report. Renewable Energy Policy Network for the 21st Century, Paris

Rotterdam (2012). Rotterdam Convention on the Prior Informed Consent Procedure for Certain Hazardous Chemicals and Pesticides in International Trade (PIC). http://www.pic.int/Countries/Parties/tabid/1072/language/en-US/Default.aspx

Stockholm (2012). Stockholm Convention on Persistent Organic Pollutants (POPs). http://www.pops.int/documents/signature/signstatus.htm

Tans, P. and Keeling, R. (2011). National Oceanic and Atmospheric Administration: Earth System Research Laboratory and Scripps Institute of Oceanography. http://www.esrl.noaa.gov/gmd/ccgg/trends/

UN (2012). "One UN" data entry point UN Data (http://data.un.org/). Last accessed 20 January 2012

UNCCD (2012). UN Convention to Combat Desertification in Those Countries Experiencing Serious Drought and/or Desertification Particularly in Africa. http://www.unccd.int/convention/ratif/doeif.php

UNCLOS (2012). UN Convention on the Law of the Sea. http://www.un.org/Depts/los/reference_files/chronological_lists_of_ratifications.htm#The United Nations Convention on the Law of the Sea

UNEP (2010). *Fisheries Subsidies, Sustainable Development and the WTO*. Earthscan, Oxford. UNEP.

UNEP (2011a). Production and Consumption of Ozone Depleting Substances under the Montreal Protocol. UNEP, Ozone Secretariat, Nairobi. http://ozone.unep.org/Data_Reporting/Data_Access/

UNEP (2011b). HFCs: A Critical Link in Protecting Climate and the Ozone Layer.

UNEP (2011c). Keeping track of our changing environment: from Rio to Rio+20 (1992 to 2012).

UNEP (2011d). Global Trends in Renewable Energy Investment 2011: Analysis in Trends and Issues in the Financing of Renewable Energy.

UNEP (2011e). *UNEP Year Book 2011: Emerging Issues in our Global Environment*. UNEP, Nairobi

UNEP (2012a). 21 Issues for the 21st Century: Result of the UNEP Foresight Process on Emerging Environmental Issues. Alcamo, J. and Leonard, S.A. (eds.).

UNEP (2012b). GEO Data Portal (http://geodata.grid.unep.ch). Last accessed 20 January 2012

UNEP-GEMS/Water (2011). GEMStat. United Nations Global Environment Monitoring System Water Programme. http://www.gemstat.org/default.aspx

UNEP-WCMC (2011). World Database on Protected Areas. UNEP World Conservation Monitoring Centre, Cambridge. http://www.wdpa.org/Statistics.aspx

UNSD (2011). Environmental indicators: Waste. United Nations Statistics Division, New York. http://unstats.un.org/unsd/environment/municipalwaste.htm

UNSD (2012a). UNSD Environmental Indicators (http://unstats.un.org/unsd/environment/qindicators.htm). Last accessed 20 January 2012

UNSD (2012b). Country Snapshots webpage (http://unstats.un.org/unsd/environment/Questionnaires/country_snapshots.htm). Last accessed 20 January 2012

UNSD (2012c). Country Files webpage (http://unstats.un.org/unsd/environment/Questionnaires/index.asp). Last accessed 20 January 2012

UNFCCC (2012). UN Framework Convention on Climate Change. http://unfccc.int/essential_background/convention/status_of_ratification/items/2631.php

US EPA (2012). Municipal Solid Waste (MSW) in the United States: Facts and Figures. US Environmental Protection Agency. http://www.epa.gov/osw/nonhaz/municipal/msw99.html.

van Donkelaar, A., Martin, R.V., Brauer, M., Kahn, R., Levy, R, Verduzco, C. and Villeneuve, P.J. (2010). Global Estimates of Ambient Fine Particulate Matter Concentrations from Satellite-Based Aerosol Optical Depth: Development and Application. *Environmental Health Perspectives*, 118(6), 847-855

Wada, Y., van Beek, L.P.H., van Kempen, C.M. , Reckman, J.W.T.M., Vasak, S., and Bierkens, M.F.P. (2010). Global depletion of groundwater resources, *Geophysical Research Letters*, 37

WGMS (2008). *Global Glacier Changes: facts and figures*. Zemp, M., Roer, I., Kääb, A., Hoelzle, M., Paul, F. and Haeberli, W. (eds.). UNEP, World Glacier Monitoring Service, Zurich.

WGMS (2011). Glacier mass balance data 1980-2010, World Glacier Monitoring Service, Zurich. http://www.wgms.ch

WHO (2006). WHO Air quality guidelines for particulate matter, ozone, nitrogen dioxide and sulfur dioxide: Global update 2005. Summary of risk assessment.

WHO (2011). Database: outdoor air pollution in cities. http://www.who.int/phe/health_topics/outdoorair/databases/en/

WHO/UNICEF (2012). Joint Monitoring Programme for Water Supply and Sanitation. http://www.wssinfo.org/

WMO (2009). 2000-2009, the warmest decade, Press release no. 869. World Meteorological Organization. http://reliefweb.int/node/336486

World Bank (2011). State and Trends of the Carbon Market 2011. Washington, D.C.

Table 1: Key environmental indicators data

Key environmental indicator	Latest year on record	World	Africa	Asia and the Pacific	Europe	Latin America and the Caribbean	North America	West Asia	Unit of measurement
Consumption of ozone-depleting substances	2010	43 292	2 559	29 971	103	5 199	2 165	3 295	million tonnes ODP
HFCs emissions - all gases	2008	651 748	2 146	237 395	140 251	14 882	255 602	1 471	gigagrams
Carbon dioxide emissions	2008	32.11	1.14	13.69	6.61	1.65	6.01	1.04	billion tonnes of CO_2
Carbon dioxide emissions per capita	2008	4.8	1.2	3.5	8.0	2.9	17.4	8.3	tonnes of CO_2 per capita
Forest net change	2005-2010	5.6	-3.4	0.5	0.9	-3.9	0.4		million hectares per year
Area protected to maintain biological diversity to surface area	2010	12.0	10.1	9.9	10.2	19.3	9.5	17.1	per cent of total territorial area
Municipal waste collection	2000-2007			271.2	537.9	130.8		20.2	million tonnes
Total water footprint per capita of national production - blue	1996-2005	167	94	181	109	110	380	345	m^3 per year per person
Total water footprint per capita of national production - green	1996-2005	1 087	1 167	780	1 259	1 924	2 689	426	m^3 per year per person
Access to sanitation	2010	61.0	39.9	57.4	90.9	80.1	100.0	78.3	per cent of total population
Number of certifications of the ISO 14001 standard	2010	251 000	1 700	131 700	103 700	7 231	5 500	1 200	number of certifications

Species trade 2010 number of wild animals (million)	
Captive bred and born specimens	321.2
Wild specimens	344.5

Renewable energy supply index 2009 (1990 = 100)	
Solar photovoltaics	86 650
Solar thermal	674
Wind	7 033
Biofuels - biogasoline and biodiesel	6 347

Primary energy supply 2009 oil equivalent (billion tonnes)	
Crude oil and feedstocks	4.10
Coal and coal products	3.30
Gas	2.54
Combustible renewables and waste	1.24
Nuclear	0.70
Hydro	0.28
Geothermal	0.06
Solar/wind/other	0.04
Total supply	12.26

MEAs 2011 number of parties	
Basel	176
Cartagena	161
CBD	193
CITES	175
CMS	116
Heritage	188
Kyoto	192
Ozone	196
Ramsar	160
Rotterdam	142
Stockholm	176
UNCCD	193
UNCLOS	162
UNFCCC	195

Acknowledgements

Year in Review

Authors:

Sarah Abdelrahim, UNEP, Nairobi, Kenya
Tessa Goverse, UNEP, Nairobi, Kenya

Reviewers:

Susanne Bech, UN-HABITAT, Nairobi, Kenya
Sophie Bonnard, UNEP, Paris, France
John Christensen, UNEP Risoe Centre on Energy, Climate and Sustainable Development, Roskilde, Denmark
Anna Donners, UNEP, Nairobi, Kenya
Robert Höft, Convention on Biological Diversity, Montreal, Canada
Ben Janse van Rensburg, Convention on International Trade in Endangered Species of Wild Fauna and Flora, Geneva, Switzerland
Sunday Leonard, UNEP, Nairobi, Kenya
Julie Marks, UNEP, Geneva, Switzerland
Richard Munang, UNEP, Nairobi, Kenya
Martina Otto, UNEP, Paris, France
Pascal Peduzzi, UNEP, Geneva, Switzerland
Mark Radka, UNEP, Paris, France
Andrea Salinas, UNEP, Panama City, Panama
John Scanlon, Convention on International Trade in Endangered Species of Wild Fauna and Flora, Geneva, Switzerland
Muralee Thummarukudy, UNEP, Geneva, Switzerland
Frank Turyatunga, UNEP, Nairobi, Kenya
Kaveh Zahedi, UNEP, Paris, France
Jinhua Zhang, UNEP, Bangkok, Thailand

The Benefits of Soil Carbon

Authors:

Reynaldo Victoria (chair), University of São Paulo, São Paulo, Brazil
Steven Banwart, University of Sheffield, Sheffield, United Kingdom
Helaina Black, James Hutton Institute, Aberdeen, United Kingdom
John Ingram, Environmental Change Institute, Oxford University Centre for the Environment, Oxford, United Kingdom
Hans Joosten, Institute of Botany and Landscape Ecology, Ernst-Moritz-Arndt-University Greifswald, Greifswald, Germany
Eleanor Milne, Colorado State University/University of Leicester, Leicester, United Kingdom
Elke Noellemeyer, Facultad de Agronomía, Universidad Nacional de La Pampa, La Pampa, Argentina

Science writer:

Yvonne Baskin, Bozeman, United States

Reviewers:

Asma Ali Abahussain, Arabian Gulf University, West Riffa, Kingdom of Bahrain
Mohammad Abido, Damascus University, Damascus, Syria
Niels Batjes, ISRIC - World Soil Information, Wageningen, the Netherlands
Martial Bernoux, Institut de Recherche pour le Développement, Montpellier, France
Zucong Cai, Nanjing Normal University, Nanjing, China
Carlos Eduardo Cerri, University of São Paulo, São Paulo, Brazil
Salif Diop, UNEP, Nairobi, Kenya
Roland Hiederer, European Commission Joint Research Center, Ispra, Italy
Jason Jabbour, UNEP, Nairobi, Kenya
Nancy Karanja, University of Nairobi, Nairobi, Kenya
Fatoumata Keita-Ouane, UNEP, Nairobi, Kenya
Rattan Lal, Ohio State University, Columbus, United States
Newton La Scala Jr., Universidade Estadual Paulista, São Paulo, Brazil
Erika Michéli, Szent István University, Godollo, Hungary
Budiman Minasny, University of Sydney, Sydney, Australia
Patrick M'mayi, UNEP, Nairobi, Kenya
Bedrich Moldan, Charles University Environment Center, Prague, Czech Republic
Luca Montanarella, European Commission Joint Research Center, Ispra, Italy
Walter Alberto Pengue, Universidad Nacional de General Sarmiento, Buenos Aires, Argentina
Jörn Scharlemann, UNEP-WCMC, Cambridge, United Kingdom
Mary Scholes, University of the Witwatersrand, Johannesburg, South Africa
Darrell Schulze, Purdue University, West Lafayette, United States
Gemma Shepherd, UNEP, Nairobi, Kenya
Steve Twomlow, UNEP, Nairobi, Kenya
Ronald Vargas Rojas, Food and Agriculture Organization of the United Nations, Rome, Italy
Ernesto F. Viglizzo, Instituto Nacional de Tecnología Agropecuaria, La Pampa, Argentina

Closing and Decommissioning Nuclear Power Reactors

Authors:

Jon Samseth (chair), SINTEF Materials and Chemistry, Norwegian University of Science and Technology, HIOA, Trondheim, Norway
Anthony Banford, University of Manchester, Manchester, United Kingdom
Borislava Batandjieva-Metcalf, Borislava Batandjieva Consultancy Services, Vienna, Austria
Marie Claire Cantone, University of Milan, Milan, Italy
Peter Lietava, Department of Radioactive Waste and Spent Fuel Management, State Office for Nuclear Safety, Prague, Czech Republic

Hooman Peimani, Energy Studies Institute, National University of Singapore, Singapore
Andrew Szilagyi, U.S. Department of Energy, Washington, DC United States

Science writer:

Fred Pearce, London, United Kingdom

Reviewers:

John Ahearne, Sigma Xi, Research Triangle Park, United States
Attila Aszódi, Institute of Nuclear Techniques, Budapest University of Technology and Economics, Budapest, Hungary
Yasmin Aziz, UNEP, Washington, DC, United States
Malcolm Crick, UN Scientific Committee on the Effects of Atomic Radiation, Vienna, Austria
Sascha Gabizon, Women in Europe for a Common Future, Munich, Germany
Bernard Goldstein, Graduate School of Public Health, University of Pittsburgh, Pittsburgh, United States
José Luis González Gómez, Empresa Nacional de Residuos Radiactivos, Madrid, Spain
Christina Hacker, Environmental Institute of Munich, Munich, Germany
Peter Kershaw, Centre for Environment, Fisheries and Aquaculture Science, Lowestoft, United Kingdom
Michele Laraia, International Atomic Energy Agency, Vienna, Austria
Sunday Leonard, UNEP, Nairobi, Kenya
Con Lyras, Australian Nuclear Science and Technology Organisation, Kirrawee, Australia
Oleg Nasvit, National Institute for Strategic Studies under the President of Ukraine, Kiev, Ukraine
Charles Negin, Project Enhancement Corporation, Germantown, United States
Hartmut Nies, International Atomic Energy Agency, Monaco
Thiagan Pather, National Nuclear Regulator, Centurion, South Africa
Nora Savage, U.S. Environmental Protection Agency, Washington, DC, United States
Ashbindu Singh, UNEP, Washington, DC, United States
Anita Street, U.S. Department of Energy, Washington, DC, United States

Key Environmental Indicators

Authors:

Márton Bálint, Budapest, Hungary
Andrea de Bono, UNEP/GRID-Europe, Geneva, Switzerland
Tessa Goverse, UNEP, Nairobi, Kenya
Jaap van Woerden, UNEP/GRID-Europe, Geneva, Switzerland

Reviewers and contributors:

Barbara Clark, European Environment Agency, Copenhagen, Denmark
Anna Donners, UNEP, Nairobi, Kenya
Hans-Martin Füssel, European Environment Agency, Copenhagen, Denmark

Kelly Hodgson, United Nations Environment Programme-Global Environment Monitoring System/Water Programme, Burlington, Canada
Robert Höft, Convention on Biological Diversity, Montreal, Canada
Eszter Horvath, UN Statistics Division, New York, United States
Rolf Luyendijk, UNICEF, New York, United States
Roberta Pignatelli, European Environment Agency, Copenhagen, Denmark
Richard Robarts, United Nations Environment Programme-Global Environment Monitoring System/Water Programme, Burlington, Canada
Reena Shah, UN Statistics Division, New York, United States
Marcos Silva, Convention on International Trade in Endangered Species of Wild Fauna and Flora, Geneva, Switzerland
Ashbindu Singh, UNEP, Washington, DC, United States
John van Aardenne, European Environment Agency, Copenhagen, Denmark
Frank van Weert, International Groundwater Resources Assessment Centre, Delft, the Netherlands
Michael Zemp, World Glacier Monitoring Service, Zurich, Switzerland

Photo credits for 2011 calendar:

Márton Bálint, Tree with branches
JJ Cadiz, Bird
Harvey Croze, Wangari Maathai
United States Geological Survey, Monitoring vessel
Ron Prendergast, Turtle eating bag
National Aeronautics and Space Administration, Cyclone
National Aeronautics and Space Administration, Planet

UNEP Year Book 2012 Production Team

Editor-in-chief:
Tessa Goverse, UNEP, Nairobi, Kenya
Project team:
Sarah Abdelrahim, Peter Gilruth, Tessa Goverse, David Kimethu, Christian Lambrechts, Brigitte Ohanga, UNEP, Nairobi, Kenya, **Márton Bálint**, Budapest, Hungary
Copy editor:
John Smith, Austin, United States
Collaborating center (emerging issues):
Véronique Plocq-Fichelet and **Susan Etienne Greenwood**, Scientific Committee on Problems of the Environment (SCOPE), Paris, France
Review editor (emerging issues):
Paul G. Risser, University of Oklahoma, Norman, United States
Graphics and images:
Márton Bálint, Budapest, Hungary, **Audrey Ringler,** UNEP, Nairobi, Kenya (cover design)
Special contributor:
Nick Nuttall, UNEP, Nairobi, Kenya

Questionnaire

You can also provide your feedback on-line at www.unep.org/yearbook/2012

Please take a few minutes to fill out this questionnaire. Your comments will help us improve future editions of the UNEP Year Book.

The UNEP Year Book 2012 presents recent scientific developments and important emerging issues in our changing environment. The UNEP Year Book is produced by the United Nations Environment Programme (UNEP), in collaboration with environmental experts around the world.

1. How clearly did each of the UNEP Year Book chapters convey the information?

	Very clearly	Clearly	Not very clearly	Not clearly at all	No opinion
Year in Review					
The Benefits of Soil Carbon					
Closing and Decommissioning Nuclear Power Reactors					
Key Environmental Indicators					

Please provide any additional comments on the content of the chapters:

2. How relevant did you find the information presented in each chapter?

	Very relevant	Relevant	Not very relevant	Not relevant at all	No opinion
Year in Review					
The Benefits of Soil Carbon					
Closing and Decommissioning Nuclear Power Reactors					
Key Environmental Indicators					

Please provide any additional comments on the information in the chapters:

3. Help us in the preparation of the next Year Book by suggesting newly emerging issues that could be considered.

4. Personal information (optional):

Type of organization:	Position:	How will you use the information from the Year Book ?	Please indicate your region:
Government	Minister/director	Private interest	West Asia
Development organization	Manager	Commercial	North America
Non-governmental/civil society	Advisor	Research/academic	Europe
Academic/research institution	Scientist/specialist	Policy making	Asia and the Pacific
International organization	Student	Education/teaching	Africa
Private sector	Journalist	Development work	Latin America and the Caribbean
Press or media	Consultant	Other (please specify):	
Other (please specify):	Other (please specify):		

Mail the completed questionnaire to: United Nations Environment Programme, P.O. Box 30552, 00100 Nairobi, Kenya or year.book@unep.org